Geometry & Colours of Meliponine Brood Cells

Emphasis on *Tetragonula*

Author: Abu Hassan Jalil
Editor: David W. Roubik

IBRA
INTERNATIONAL BEE
RESEARCH ASSOCIATION

Jointly published by: The International Bee Research Association,
a Company Limited by Guarantee, 1, Agincourt Street, Monmouth, NP25 3DZ (UK) &
Northern Bee Books, Scout Bottom Farm, Mytholmroyd, Hebden Bridge HX7 SJS (UK).

Obtainable from:
www.ibra.org.uk & www.northernbeebooks.co.uk

ISBN: 978-1-913811-09-9

Artwork by DM Design & Print
IBRA Proof Editor - Stuart A. Roberts
Cover image: Creative Commons, Attribution-Share Alike 4.0 International

Title: Geometry and Colours of Meliponine Brood Cells – Part 1

Subtitle: Preliminary Emphasis on Genus *Tetragonula*

Synopsis:

This literature is for enthusiasts who have difficulty remembering scientific names but can recognize geometric shapes. A preliminary look at what is termed as the most difficult genus in the Indo-Malaya stingless bee group to identify and differentiate between species by morphology. This book will take the readers on a journey to explore the various aspects of *Tetragonula* brood structure and form across the Indian Subcontinent, South East Asia, Sunda, Wallacea and the Sahul regions.

A study on brood cell shapes of the different stages of the cell build-up. From newly operculated cells, larval brood cells to pre-pupal and cocoons. Queen cells are scrutinized in size, proportion, and shape. The colour of the cells with the bee's anatomical colours and host cavity is examined.

There is a schematic impression of the development and a growth sequence of clusters, clumps, semi-combs, full layered and spiral combs. A brief glimpse is provided of the brood of some other genera like *Austroplebeia, Geniotrigona, Heterotrigona (Platytrigona), Homotrigona (Lophotrigona), Lisotrigona,* and *Wallacetrigona* are given as comparisons.

Where available, the nest and brood structure are associated with its nest entrance, which I hope will make it easier to recognize a particular species.

Acknowledgements for Parts 1 & 2

I thank Dr. Anne Dollin and Prof Siti Salmah Idrus for encouraging me in scrutinizing brood structures to learn the differences in Meliponine brood cell formation. I am also fully grateful to Dr. David Roubik, Prof Dr. Deborah Smith, Prof Dr. Cleofas Cervancia and Dr. Hans Bänziger, who has been an integral part of inspiration and encouragement in my quest for knowledge in various aspects of stingless bees and Meliponiculture.

A debt of gratitude to Dr. Shashidhar Viraktamath of India and Dr. Pham Hong Thai of Vietnam for their literary contributions and inputs on brood cell data on species in their respective regions. I also appreciate the many stingless beekeepers throughout the Indo-Malayan and Australasian regions, who contributed images of the Tetragonula brood for this study.

Acknowledgements for Part 3

I am deeply grateful to Dr. Claus Rasmussen for his advice and help on most taxa's etymology. Without his advice and new key on the genera and subgenera of Indo-Malaya, Australasia and Papuasia, this study may not have been possible to complete.

Much appreciation is for Dr. Anne Dollin for her advice and helps to provide much information on Australian stingless bees in the Australasian taxa.

My gratitude goes to Dr. Michael S. Engel of the University of Kansas & AMNH (1998–2000) for providing updated information on the latest taxa for Meliponini, Apidae Hymenoptera and other corrections herein.

This author wishes to thank Dr. Shashidhar Viraktamath of the Department of Entomology, University of Agricultural Sciences, Bengaluru, Karnataka, INDIA, for providing his etymology. "Two New Species of *Lisotrigona* Moure" was published in 2017.

Foreword

This book presents a refreshing and intriguing approach to stingless bee species differentiation and recognition. Mr. Abu Hassan Jalil has explored a new perspective of the Indo-Malayan stingless bee clade's most difficult to identify and obscure genus. Over the years, many taxonomists have looked closely at the *Tetragonula* genus and streamlined the large variety of species into synonyms, yet still end up with a long list of species that are extremely difficult to differentiate morphologically. Some have resorted to DNA sequencing to differentiate them, but these methods do not help the common beekeepers, who are faced with the difficulty of segregating the honey that each species produces.

The author has provided in-depth analyses and observations on the nest structures and brood cells for over a decade. This period was attested by some senior scientists acquainted with his studies. Taxonomists and entomologists seldom get the opportunity to scrutinize the innards of Meliponine nests because the nests' build-up is usually within cavities of rock crevices or tree hollows. The profession of the beekeeper cum amateur entomologist, as the author claims to be, gets to examine and scrutinize the makings of a nested structure in their hives.

The classification developed and proposed by the author seems easier for the beekeepers, although the species identification may need further verification. He has, however, indicated that the species mentioned remained suspect, even though some were verified by the late C. D. Michener in Thailand.

Overall, this book represents a new way of having a supplementary criterion for identifying stingless bee species. We owe a deep sense of gratitude to Mr. Abu Hassan Jalil for graciously sharing his extensive knowledge and practical experiences and bringing out this timely publication. This book is an important milestone in the world of stingless bees and is thoroughly edited by Dr. David W. Roubik, a research entomologist from the Smithsonian Tropical Research Institute, Panama. I hope this book, published by IBRA, will be widely used by beekeepers, researchers, extension workers, museum scientists, policy makers, consumers and students to advance Meliponiculture globally.

Ravindra C. Joshi

Pacific Coordinator, SAFE Network, Indonesia

Visiting Professor, Central Bicol State University of Agriculture, Philippines.

Contents

Preface

Some time ago, I was in anguish because the Indo-Malayan beekeepers were uninformed or unsure of which type of stingless bee they were keeping. It sets me wondering what species of *Tetragonula* honey they were selling. As it turns out, almost all Indian beekeepers were producing *T. iridipennis* honey, and all the Javan beekeepers were producing *laeviceps* honey. "Cannot be", I thought. There is such a diverse distribution of *Tetragonula* and yet only one species in each region? In many other regions, e.g., in Brunei and the rest of Borneo, it was either *T. fuscobalteata* or *Tetragonula* sp. In the Philippines, every stingless bee other than *T. biroi* was called 'monggo-monggo'. In other words, all of the *Tetragonula* honey throughout was just a mix of honey of many different species. 'Iridipennis' means iridescent wings. As all stingless bee wings glimmer iridescently in one angle or another, every stingless bee is now *T. iridipennis.*

In Borneo, a rumour went around that an old lady's brain tumour ailment was cured by taking *fuscobalteata* honey. It seems that this was the easiest stingless bee to obtain as it is prolific in colony multiplication and most recognizable as the tiniest stingless bee. In Brunei, *T. fuscobalteata* honey is favoured for its sweetness and sold in Brunei dollars B$ 15.00 per 80ml bottle. So, to the neighbouring Kalimantan beekeepers, that's good money, and so, *T. fuscobalteata* was much sought after. Then in Lombok, which only had *T. sapiens* and *T. clypearis* (tiny bees with equally sweet honey) and their honey were quickly taken up by buyers in Java.

Java does not seem (i.e., amongst stingless beekeepers) to have any stingless bee other than *T. laeviceps*; it does. It is just that many stingless bees honey producers don't know how to differentiate the species. So, everything is lumped as one that is *T. laeviceps* honey. The predicament is that the *Tetragonula* honey in those regions mentioned does not carry any credible determination of actual bee origin.

I decided I may do something about this. I began to compile images of any *Tetragonula* brood cells and compare them. I was inclined to group the compilation into regions, i.e., the Indian Subcontinent, Sunda shelf, Maritime Wallacea and the Sahul shelf.

Besides this, I was also looking for *Tetragonula* brood cell configuration and structure images during this restricted movement period of the current Covid19 Pandemic. I sifted through my past years' image library that started with 24 images. I requested more from friends and beekeeping acquaintances while I began to see some consistent patterns in the clustered brood type. The response was encouraging, and I managed to classify the brood cell shapes geometrically. The initial image database grew into an image bank, and I decided to break the whole pile into regions and came up with the following:

1. Indian Subcontinent; 2. Mainland SEA; 3. Sunda Shelf; 4. Wallacea & 5. Sahul. I then shared different sets with email loops and social media interest groups. The sets were segregated with different regionally significant relevance for corresponding enthusiasts and academicians of the related region.

Goodness gracious me, the response and feedback are overwhelming, and the image database has grown to almost 200 images of clustered and semi-comb brood cell structures. I have had to break the sets up further and add a few categories.

1. India has Nagaland layered brood images moved to a new set -*Tetragonula* layered and semi-combs. A conspicuously bicolorous layered brood is similar to *T.* aff. *sarawakensis* and seems to appear across the region. from Nagaland to Cambodia to Vietnam to Sabah. North Borneo, what more, Sarawak & Singapore. If not for the image database, I would not have guessed that this species (*T. sarawakensis*) may be this widespread.

2. Mainland SEA is split into 2a. Peninsula Malaysia & Thailand with 2b. Cambodia & Vietnam, here I include *Lisotrigona furva*. We want to compare *Tetragonula (Tetragonilla) atripes* with *Tetragonula (Tetragonilla) collina* in Thailand, Vietnam, Sumatra, West Malaysia and East Malaysia.

3. The Sunda shelf is split into 3a. Sumatra & Java and 3b. Borneo (i.e., East Malaysia & Kalimantan). Here, I aim to include *Borneotrigona hobbyi;* after all, only one sp.

4. Wallacea – here, it includes Indonesian Wallacea and Oceanic Philippines. I move the *T.* nr. *biroi* of Sulawesi, *T.* nr. *biroi* of the Philippines and *T. biroi* of West Papua into a comparative set - the new set -*Tetragonula* layered and semi-combs.

5, Sahul - a split here sees 5a, the regular cluster types, 5b. *Austroplebeia* spp. and the *T. carbonaria, T. hockingsi* and *T. davenporti* moved to the new set -*Tetragonula* layered and semi-combs.

So, in total, I have nine sets according to regions and types. I will share these new sets as I update them, most likely in future volumes or series. The reader may correspond through emails (abuhsn@gmail.com) to let me know which region is of interest or which images one would like to contribute to the database. I have not included any *Tetragonula (Flavotetragonula)* spp. as yet. I do not have brood cell data on this new Subgenus. (See Appendix).

Initially, I was intrigued by the diversity of the *Tetragonula* genus, and many told me that it is a very difficult genus to identify the species. The morphological descriptions were in dichotomous keys, and they were in batches of bees found in certain regions at specific times. When one picks out the key for all *Tetragonula* only, it gets very, very confusing (maybe not for taxonomists but to the meliponiculturists, it is mind-boggling).

Anyhow, I was then inspired to work on the *Tetragonula* brood cells when I read three works:

1) Biometric Studies of the Stingless Bee *Tetragonula laeviceps* Complex (Apidae: Meliponini) from Northern Thailand, Chuttong & Burgett 2017

2) Colony Characterization and Amplicon Sequencing of Stingless Bees (*Tetragonula biroi*), Baja & Mostoles 2017

3) Nest architecture, colony composition and morphology of *Tetragonula (Tetragonula) drescheri* Schwarz (Hymenoptera: Apidae: Meliponini), Salmah & Yasmi 2017

All three works look at the brood cells and their cell peculiarities in detail besides other aspects. I started looking at *Tetragonula* spp. brood, and so far, I have seen only *T. carbonaria, T. biroi,* and *T.* cf. *sarawakensis* has a full comb brood arrangement. (Other than the species complex of *T. laeviceps*). I have not gotten to view (what I would presume resembling) *T. penangensis*. Others like *T. drescheri, T. hockingsi* or *T. davenporti* have semi-comb or discontinuous segmented comb. In my image bank of brood cells, images from Vietnam have many examples which have both discontinuous segmented comb, semi-comb or mini wafer combs combined with clumps of clustered cells in the same hive at the same time.

I cannot place or compare them with any[1]; I merely call them undetermined species on their respective brood cell shapes.

Editor's Notes

[1] This describes and explains most of the problem here. These things vary. One needs to understand the range of variation within species, then ask the question whether the different species are distinguishable. Sakagami did this kind of thing with morphology using, conveniently, three axes or dimensions- x, y and z. One could do this with the brood cell arrangements with several more, and with the cell size and shape, which is not independent of the brood cell arrangement. Some actual statistical tests could be applied to this kind of thing- which is a problem that comes up all the time in the study of morphology.

Henceforth, this write-up will attempt to classify those brood cell shapes and describe the types of cells, be they newly operculated, larval, or mature prepupae and imagos[2] in their cells. The brood form and structure are also categorized so that beekeepers and meliponiculturists can identify the honey they produce. It is my hope that this contribution may open a doorway for more than half of total bee enthusiasts to appreciate the diversity of stingless bees in this region.

[2] imago is an adult- it has gone through metamorphosis. A cocoon is spun by a mature larva. Then it is a resting prepupa for a while, and then it pupates and becomes a pupa, and the pupa matures to an adult bee or imago, which then emerges as a callow adult from its cocoon. The cocoon is ALL THAT THERE IS, after the ORIGINAL OUTER LAYER OF THE BROOD CELL, THE CERUMEN, IS TAKEN AWAY, WHICH HAPPENS SHORTLY AFTER THE COCOON IS SPUN BY THE MATURE LARVA.

Part 1

Abstract to Part 1

The most problematic genus in recognizing its species in the Indomalayan and Australasian stingless bee realm is Genus *Tetragonula* Moure, 1961. With its two subgenera i) *Tetragonilla* Moure, 1961 having four species, and ii) *Tetragonula* Moure, 1961 having 31 species (Rasmussen, Thomas & Engel 2017). This chapter scrutinizes the brood cell shapes of the *Tetragonula* subgenus cluster type or semi-comb type. The full layered or spiral combs are also examined and categorized accordingly. They are grouped into five main geographical regions, 1. Indian Subcontinent; 2. Mainland SEA; 3. Sunda Shelf; 4. Maritime Wallacea and 5. Sahul Shelf. Two subregions are not fully explored, i.e., the southern tip of India with Sri Lanka and the Papuasia region, which includes both Wallacea and Sahul.

The shape classifications and the related species mentioned are hypothetical and not applied for species determination. The brood cell and forms examined are from macro photographs and micrographs, and the PowerPoint sizing feature determines the cell height to cell width ratio. They are classified into six basic shapes: Ellipsoid. Cylindroid, Ovaloid, Prolate or Oblate Spheroid and Cuboid, each defined herein. The types of brood formation are detailed, and the combination of brood cell shape mixes are elaborated.

Keywords: Brood cell shapes, Meliponine, Meliponiculture, Stingless bee Honey, Brood form, Nest structure, Spiral brood-comb.

Chapter 1

1.0 Introduction

The year 2019 brought me to the 1st Australian Meliponine Conference. While I was visiting Meliponaries observing bee brood in Malaysia, Thailand and Indonesia. Now I had the opportunity to observe *Tetragonula* brood in Australia. I spent a week with Robert Luttrell, and he introduced me to "Bee Celebs" Peter Davenport, John Klumpp, the late Alan Beil and Nick Powell. AAA 2020 brought me to the Philippines, and this time I had the opportunity to travel to a few Islands. Together with Dr. Roubik, we observed some *Tetragonula* nest structures kept by the local beekeepers in the Philippines.

While in Australia, I made illustrations of the unique native stingless bee and looked into the morphology. The *Austroplebeia* were great, but the Sahul *Tetragonula* was more intriguing because of the familiarity of the genus to my Sunda origin. Dr. Anne Dollin told me that one could only tell them apart from the brood structure and forms, especially *T. hockingsi, T. davenporti* & *T. mellipes*. Prof Deborah Smith believes that it is very difficult to differentiate the Indo-Malayan *Tetragonula* other than body size. How does one accurately and consistently ascertain the body size when the dead stingless bee shrinks under a microscope, and to top it, the tergites are telescopically variable.

Excluding a full layered or spiralling comb-type brood, there are six basic brood cell shapes and various types constantly having conspicuous connectors, some thick, some fine or just one or two attached filaments. The species classification is hypothetical, but this compilation ascertains brood shape categories beyond verifying the species identity.

Across the board, there were similarities of brood cell shape categories for bees that overlapped existing inter regions. E.g., *T. iridipennis* appears in India and Sunda & Wallacea. *T. testaceitarsis* & *T. valdezi* appears in Sunda and Wallacea. *T. sapiens* and *T. clypearis* appear in Wallacea as well as Sahul. *T. laeviceps* & *T. fuscobalteata* appears in almost all the regions.

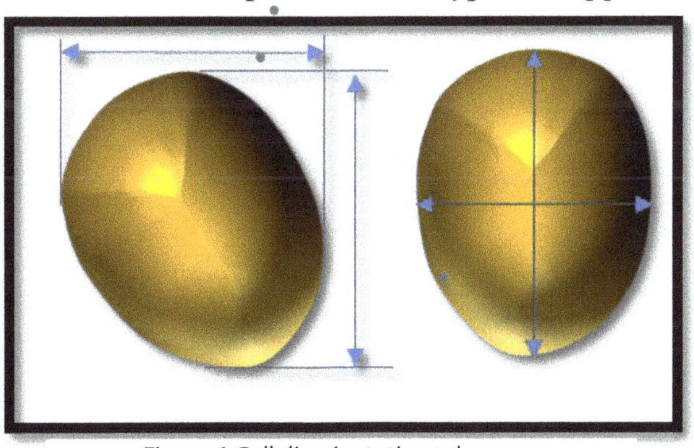

Figure 1 Cell disorientation tolerances

As I looked at clustered cells, graduated brood and semi-combs, the random orientation posed an uphill task of measuring accurately, and the variations would be too drastic. With less complicated methods in mind, I decided to work with ratios. I used the sizing feature in Microsoft PowerPoint to obtain the height(h) / width(w) ratio, seemingly most appropriate. Where cell connectors (c) appear, the ratio of c/w seems prudent.

I have inferred so far browsing through the image database I mentioned previously.

I found that six basic shapes appear consistently among the brood cells of stingless bees in Indo Malaya. In keeping with the tradition of Latin and Greek etymology of the geometric terms of Trigonum and Tetragon used in Meliponine vocabulary, I used geometric terms to categorize the brood cell shapes.

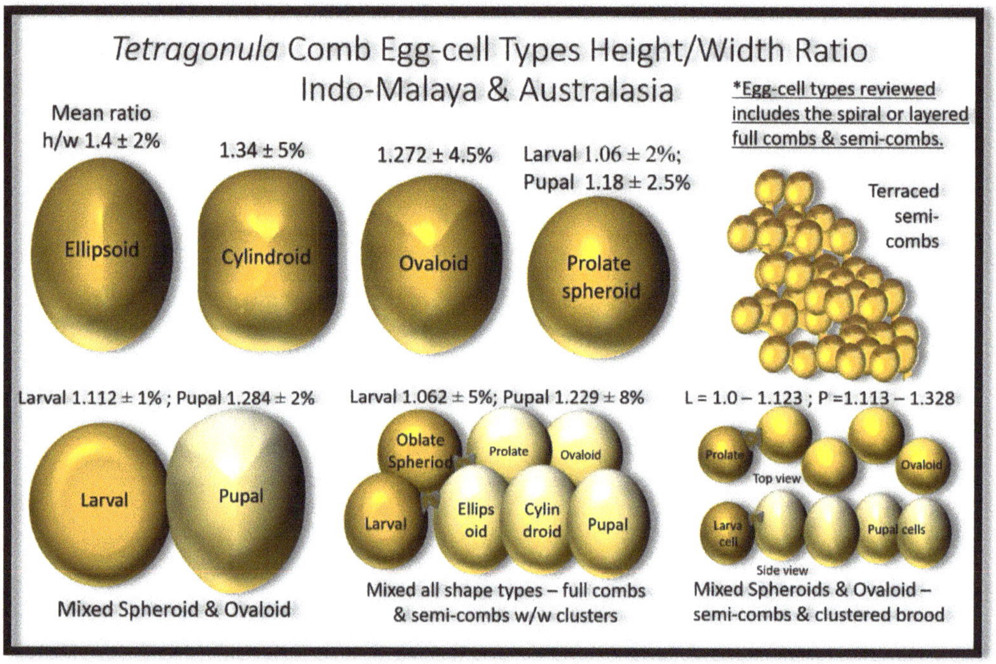

Figure 2 Tetragonula Clustered Brood-cell Types: Height/Width Ratio.

Figure 3 Progression of Clustered Type Brood Cell Stages

1.1. Progression of Clustered Type Brood Cell Stages

There are six stages of progression.

A. Initial Empty Cell
   - i. Provisioning open cell
   - ii. Preoviposition
   - iii. Pre-Capping

B. Newly capped cell (Operculated)

C. Hatched egg within - Larval cell

D. Metamorphosis - pre-pupal cell

E. Cocoon stage - Pupal cell

F. Remnant cell shell - Post callow emergence.

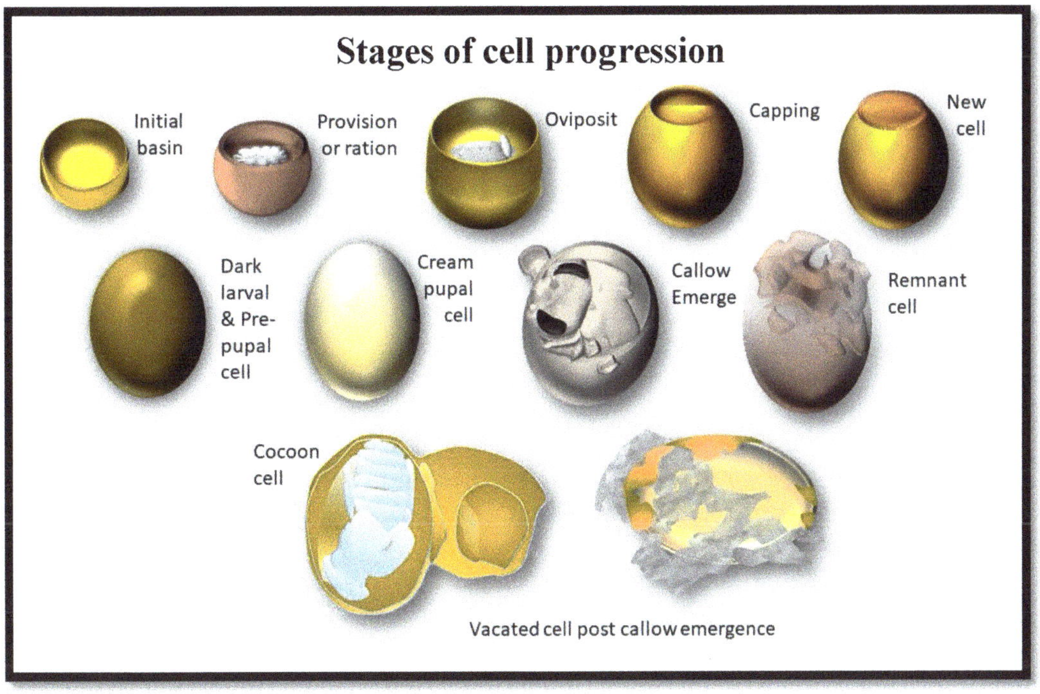

Figure 4 Stages of Cell Incipiency.

I have asked this question in my mind while doing the illustrations when examining the brood cell development sequences: When and why do some callows emerge at the top cap while some break through the sidewall? The cell-to-cell connectors and the brood cell clusters or semi-combs to involucrum or batumen connectors play a role in this mode of callow-birth (if there is a term for this). The position of the connectors relative to each cell is not consistent and will vary depending on the cavity the brood chamber occupies. Feral colonies in tree hollows, bamboo cavities and culture of brood in hive boxes all differ greatly in cell connectors and position of the connectors. I started taking the connector's width (c) to cell width (w) ratio (c/w) just for comparison. When the connectors attach at the cap's brim, nurse bees can assist in uncapping for easier exit. In hollow tubes like bamboo or, the connectors tend to be attached end to end, i.e., top of the cap to the base of the adjacent cell. In this instance, the exit is bursting the sidewall. Those in hive boxes have leeway to cluster in horizontal layers, and the cells have

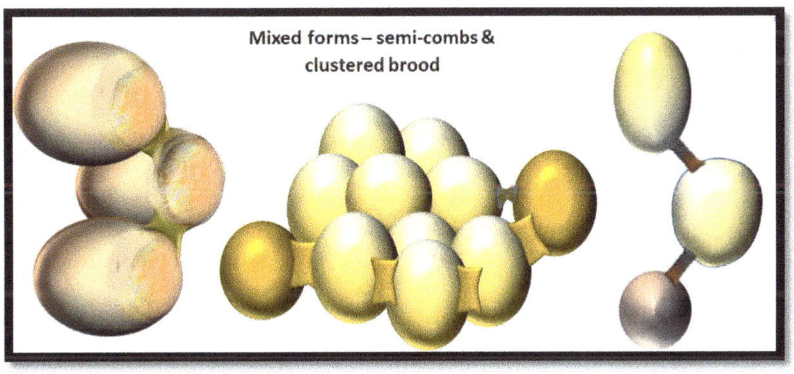

Figure 5 connectors width(c) to cell width(w) ratio (c/w) just for comparison.

vertical flat connectors attached to the sides. Semi-comb groups have cells attached to each other, and each group then has strong connectors to other groups. I am still exploring variations in the connectors and each scenario's peculiarity. I may include this in the next series of this book.

Dr. Roubik suggested 'ovoid' instead of 'ovaloid'. All these shape terms are part of palynology and pollen forms. I was torn between the two initially. Then I looked up the difference - As nouns, "the difference between 'ovoid' and 'oval' is that ovoid is 'something that is oval' while oval is a shape rather like an 'egg' or an ellipse." I chose ovaloid as it is about eggs (the top and bottom hemispheres are asymmetric). Thanks to Dr. Roubik for pointing out similarities in pollen grain shape terms in Palynology. "It makes sense. Not exactly oval but 'egg-shaped'. I had never seen that only 'egg shaped', and we have no such pollen grains."

I appreciate Dr. Shashidhar Varaktamath for sharing the article doi:10.1111/nyas.14680 excerpt: "As the main parameter in oomorphology, the shape of a bird's egg... can be done using four geometric figures: sphere, ellipsoid, ovoid, and pyriform." (I think oomorphology is about morphology in Oology - a branch of ornithology studying bird eggs, nests and breeding behaviour.) So, concerning my classification's oomorpholoy article, I further detail the spheres into Oblate and Prolate spheroids. And pyriform does not apply. The six basic shapes that I use are explained in subsequent chapters. I have always been interested in Mimetic geometry, in the works of Ibnu Sina (Avicena) and Fibonacci, henceforth the geometry of nature.

Coming from an egg myself, I learned about the human egg early (pre-adolescent). We learn in religious studies from the Holy Quran, Surah # 96 "THE BLOOD CLOTS - AL-ALAQ" (In the Name of Allah, The Beneficent, The Merciful.) "Read in the name of your Lord who created, created man... from a dried drop of blood which sticks to the womb: a cheap and unsophisticated substance." Translated from International Islamic University Malaysia (https://www.iium.edu.my/deed/quran/in_the_shade_of_quran/s096.html) forth my background knowledge on the Human egg.

In the same context, I study the formation of the Meliponine brood cell. The cell or "womb" starts as a basin to receive provisioning and oviposition. The workers then build up the walls parabolically (cross-section). All done, the workers complete the top portion with a hyperbolic dome, which we call the cap or operculum (as in Malacology, meaning a secreted plate that closes the aperture of a gastropod mollusc's shell when the animal is retracted).

Terms and definitions Drummond et al., 1999.

The terms and definitions used in this paper are taken from Michener (1974), Sakagami and Zucchi (1974), Sakagami (1982), Roubik (1992), Sommeijer and Bruijn (1992), Sommeijer and Van Buren (1992), and Drumond et al. (1997).

Cell capping

This phase is usually divided into subphases. The pre-operculation subphase (sp.) is the interval between queen oviposition and the arrival of the first worker, which commences closing the cell. The rotation subphase (sr) is when the worker, having the tip of her abdomen in the cell, rotates within the cell and begins to close it by manipulating the collar with both mandibles and legs. The side work subphase (ss) is when the worker withdraws from the cell, and the work is then completed by the same bee or another bee or bees working from the side of the cell.

Cell provisioning and oviposition process (POP)

Sakagami, Zucchi and co-workers categorized the daily reproduction cycle into phases and subphases. Each behavioural sequence necessarily involves a provisioning phase, an oviposition phase and an operculation phase. Together, these phases are collectively known as the cell provisioning and oviposition process or POP. Depending on the species, this process may comprise constructing and provisioning a single brood cell or several cells simultaneously.

Clusters

This form is one of the several patterns of brood cell organization. In the case of brood cells organized in clusters, cells are built into an irregular side-to-side array in which no wall is shared by adjacent cells.

Collared cell

A newly constructed brood cell is complete in nearly all species when it shows a small neck called a collar.

Food provisioning

When larval food is discharged into several cells concurrently, food provisioning is regarded as 'synchronous. If cells are provisioned in turn, this is termed 'successive' provisioning.

Larval food

This substance is a soup-like mixture of pollen, honey and glandular secretion.

Post-provisioning phase (ps)

This phase is from the end of the provisioning phase to the beginning of the queen oviposition.

Pre-provisioning phase (pr)

This phase is the interval between the Queen's arrival in front of a brood cell and the first drop of larval food discharge into this cell.

Provisioning phase (p)

This phase runs from the first droplet of food into the brood cell to the end of the last one. Each provisioning phase involves several discharging acts performed by different bees (and sometimes by the same bee) successively.

Successive cell construction

Individual cells are started asynchronously so that cells in varying stages of development are seen simultaneously.

Chapter 2

2.0 Brood Cell Shape Classification

2.1 Brood-cell Shapes in Clustered *Tetragonula* spp. Brood Structures

The essential basic shapes observed were Ellipsoid, Ovaloid, Cylindroid, Oblate spheroid, Prolate spheroid and Cuboid [Cuboid - Solely found to occur in Queen cells. A cube with rounded corners. It may have two or more if its side flattened depending on its position in a comb (Figure 20 Cuboid Queen cells & Figure 21 Cuboid Queen cell shape; Cuboid definition - more or less cubic. This is the sixth variation that I have observed (seldom occurs, but it is very conspicuous))]. Other variations are pairing of any two or *mixed* combination of three. (See Figure 6.)

(*Note: I do not use the term Globoid because, in Botany, globoid cells can mean a plant cell in an irregular round shape.)

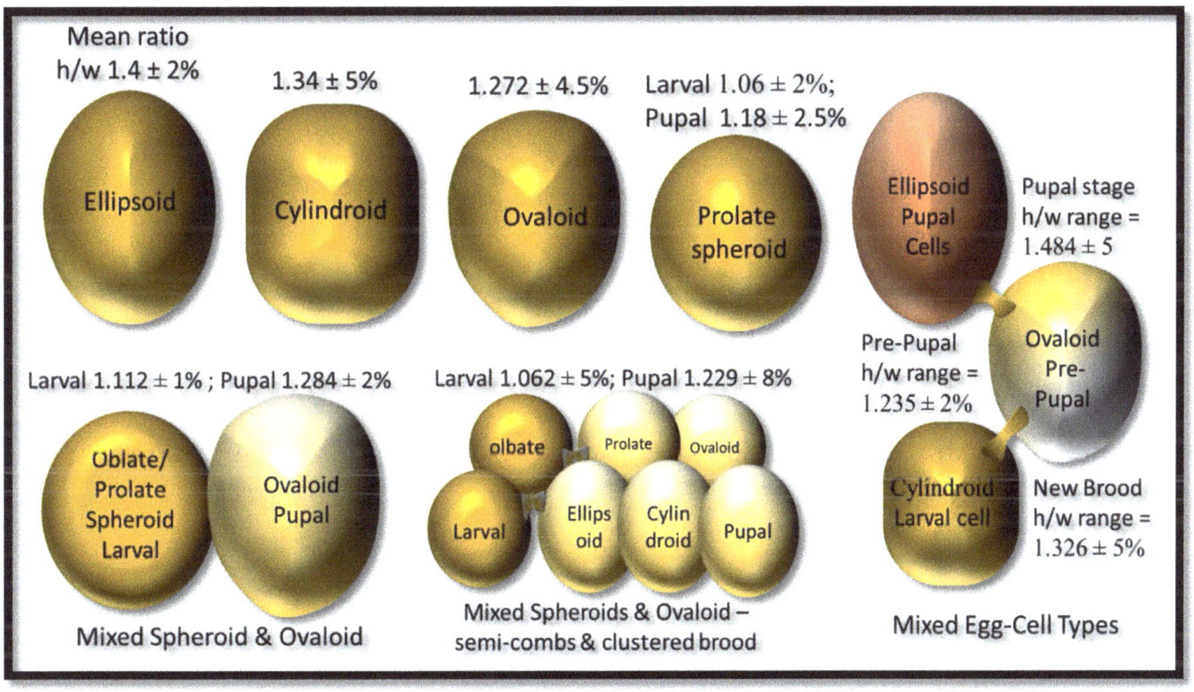

Figure 6 Tetragonula Clustered Brood-cell Types Height/Width Ratio

These terms were taken from Geometry, and the definitions herewith are in geometrical terms.

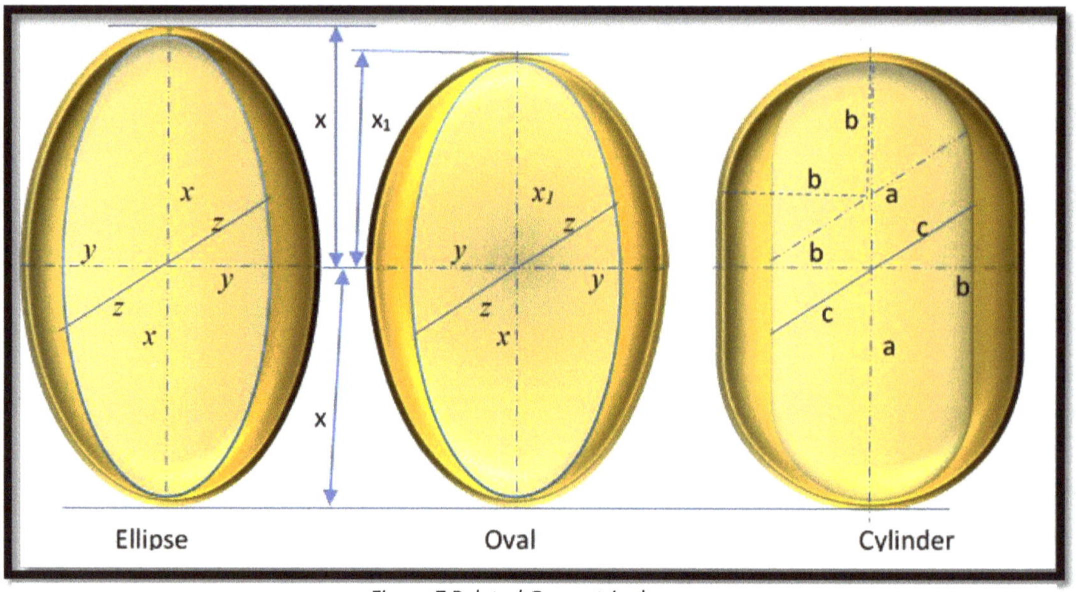

Figure 7 Related Geometric shapes

I. **Ellipsoidal definition:** Ellipsoid, the closed surface of which all plane cross-sections are either ellipses or circles. An ellipsoid is symmetrical, about three perpendicular axes intersecting at the centre. The 3D shape is a solid formed by rotating an ellipse around either axis:

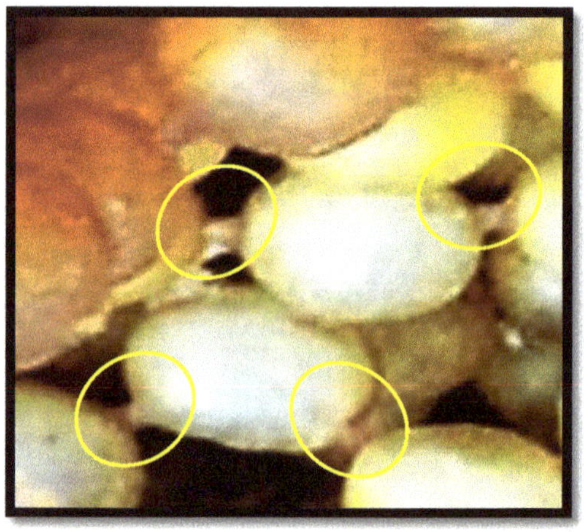

Figure 8 ellipsoid cells are connected end to end rather than side

Ellipsoidal meaning in brood cell shape:

its plane sections are ellipses or circles; in this instance, the longest vertical axis is taken upright.

Ellipsoidal connectors: As far as connectors go, ellipsoid cells are connected end to end rather than side to side in many instances. Where the connections are end to end, emergence is at the sidewall. On a broad stroke, the cell height to width ratio ranges

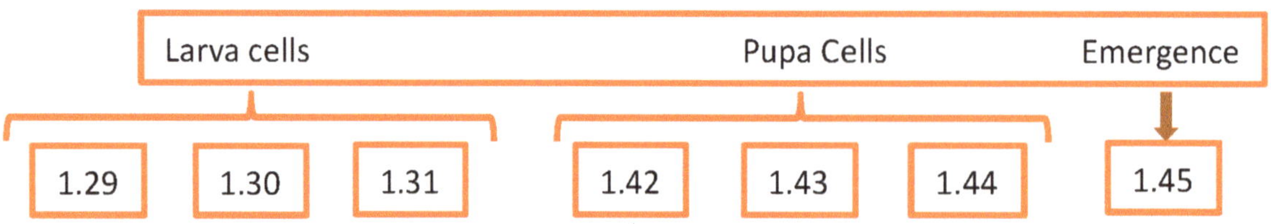

Figure 9 Relative growth of Ellipsoid shaped cells based on h/w ratio

Figure 11 Vacated cell by emerged pupa

Figure 11 cell wall side eruption

approximately from 1.3 to 1.5. This shape is evident in the medium to large body size of *Tetragonula* species. An example of pupa emergence is an h/w ratio of 1.45 to 1.51. (See image above)

II. Ovaloid

Ovaloid definition: A three-dimensional geometric figure or relating to an oval; approximately oval or ovoid - An oval (from Latin ovum, "egg") is a closed curve in a plane that resembles the outline of an eggshell.

Figure 12 A nurse bee assisting the uncapping. The rightmost cell shows a callow pushing out from the corner of the cell cap (circled in red).

Ovaloid meaning in brood cell shape: brood cell-shaped with a larger radius at the cap. The upper hemisphere is asymmetric to the lower hemisphere, and this asymmetry may have been caused during the cell cap closure (operculation). Appearing like a flattened or depressed cell cap dome (operculum).

Figure 13 cells are connected at the upper portion of the cell but below the cap line.

Ovaloid connectors: In ovaloid cells brood structure, the cells are connected at the upper portion of the cell but below the cap line, and it leaves the top dome to be easily uncapped for callow emergence.

III. Cylindroid

Cylindroid definition: resembling a cylinder with equal and parallel elliptical bases and an elliptical cross-section.

Cylindroid meaning in brood cell shape:

At a glance, it looks like a gas tank cylinder. Occurs in tight, compact clusters and semi-combs as the sides are pressed together and generally in close quarters, making them parallel. The height variation can be great and may occasionally have a bigger cell height to cell width ratio than the ovaloid h/w ratio.

Figure 14 Propane gas cylinder

Cylindroid connectors: cylindroid cell brood are random, and for semi-comb brood, the emergence is random without any set orientation. An example image of the semi-comb brood of the *Tetragonula davenporti* where many callows emerge almost simultaneously, but in many directions and different orientations. The position and orientation of the connectors influence their point of emergence. The groups of about 15 to 20 cells in small comb layers are at the sides or where the small combs connect at different levels have connectors angled to one another.

Figure 15 Tetragonula davenporti brood cells with emerging callows

This inconsistency is also evident in the semi-comb brood of *T. pagdeni*, where the connectors are all over the place. However, the *T. pagdeni* brood cell shape is classified as ovaloid. The point of mention here is that connectors in semi-comb brood have inconsistent connectors.

Figure 16 T. pagdeni brood cells with inconsistent connectors.

IV. Oblate Spheroid

Spheroid definition: a sphere-like but not perfectly spherical body; a solid generated by a half-revolution of an ellipse about its major axis (prolate spheroid) or minor axis (an oblate spheroid). A spheroid has circular symmetry.

Oblate spheroid meaning in brood cell shape:

An oblate spheroid is the body of revolution formed when an ellipse with minor axis dimension (a) and major axis dimension (b) is rotated about its minor axis; (of a spheroid) flattened at the poles. Generally, the h/w ratio is less than one.

The apparent flatness is due to the radius of the cap dome being larger than the base radius.

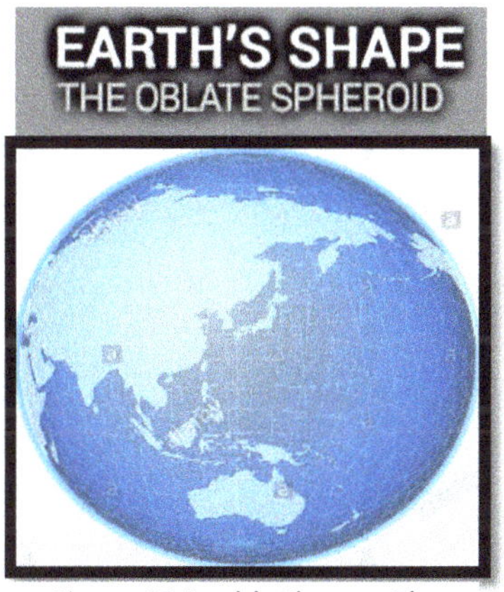

Figure 17 Earth's Shape – The Oblate Spheroid

Figure 18 T. minangkabau show short fine connectors at its "Equator"

This image example of *T. minangkabau* shows short fine connectors at the "Equator" of the oblate spheroid brood cells. Even though the connectors are multi-directional, they do not appear to impede callow emergence through the cap dome. The brood is randomly clustered with no distinct size or colour tone differences between the larval and pupal cells.

V. Prolate spheroid

Prolate spheroid meaning in brood cell shape:

An ellipsoid of revolution generated by revolving an ellipse about its major axis; a quadric surface obtained by rotating an ellipse about one of its principal axes; in other words, an ellipsoid with two halves fronts and back or top and bottom; (of a spheroid) lengthened in the direction of polar diameter. Resembles a rugby ball standing on its end but has a shorter vertical axis. The caps may appear to have a depressed cap., as in fresh larval brood. The caps inevitably will dome up as they pupate.

Stages of Prolate Spheroid shape brood cell

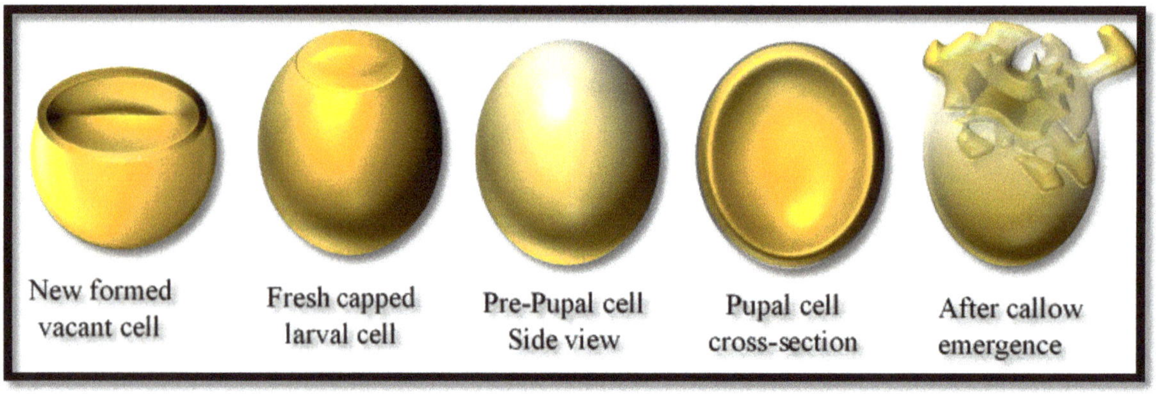

Figure 19 Stages of Prolate Spheroid shape brood cell

VI. **Cuboid definition -** more or less cubic. - This is the sixth variation that I have observed (seldom occurs, but when it does, it is very conspicuous)

Cuboid, meaning in brood cell shape: The cell is generally much larger than the other brood cells, presumably queen cells. It looks like a cube with rounded corners. Some are like cylindroids, but the height is equal to the width, i.e., h/w approx.= 1. The h/w ratio ranges from 0.988 to 1.09 and averages at 1.0467. The Cuboid cells are the extra-large ones in pale cream colour. The pupal cells in semi-combs are Cylindroid, and larval cells (not in view) are Ovaloid in separate clusters. Photo (*Figure 21*) is a terraced semi-comb brood in a box hive corner in Vietnam. Reference video internet site: https://www.youtube.com/watch?v=8e5tquxiLMc

Figure 21 Cuboid Queen cells

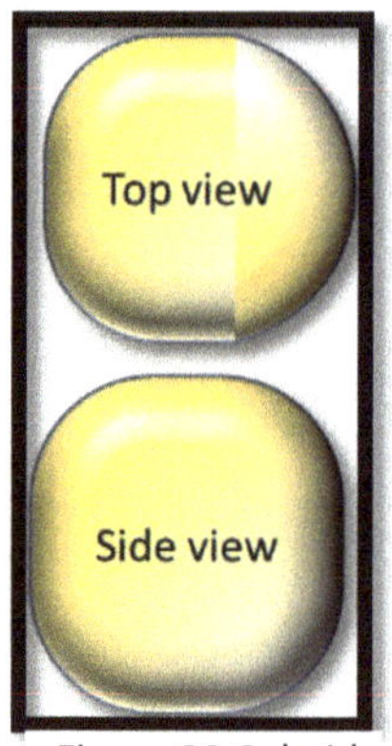

Figure 20 Cuboid Queen cell shape

The side profile of the cell is squarish with rounded corners. Sidewalls are three-sided flat (See *Figure 20*) when the queen cell is sandwiched between pupal cells at the perimeter of the semi-comb or two-sided flat when they are at the corners.

Chapter 3

3.0 Combination of Shapes

VII. Mixed Spheroid & Ovaloid Cells Ω Indiscriminate connectors

Description: In this instance, the Larval brood and the pupal cells are usually in separate clusters, and the larva clusters are much darker prolate spheroid while the pupa cells are ovaloid and very much lighter in colour tone.

Connectors: Connectors are indiscriminate, and the thickness range varies with the species involved. They may be thicker in the larval stage and diminish as they grow onto the pupal stage. In some, like *T.*cf. *testaceitarsis* or sp. they may appear like thick chain links.

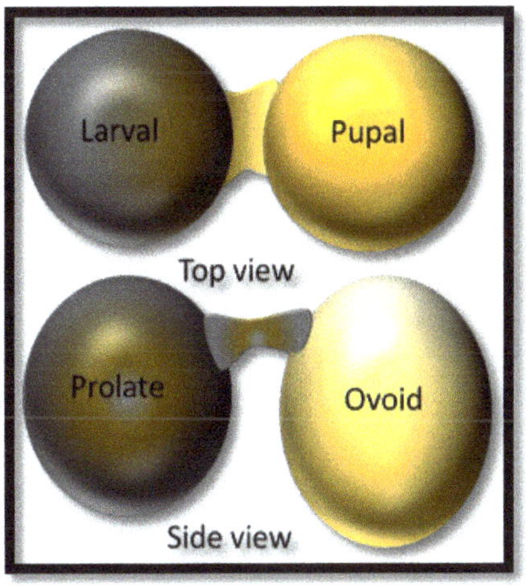

Figure 22 Mixed Spheroid & Ovaloid Cells

Figure 23 Indiscriminate connectors in T. sapiens or sp

An example image of the indiscriminate connectors in *T. cf sapiens* or sp. Connector thickness / larval cell width (c/w) of all the nests observed has a mean ratio of 0.2685. The *T. cf sapiens* has Prolate Spheroid Larval brood but Ovaloid Pupal cells

However, in *T.* cf. *testaceitarsis* or sp. (image below), the connector thickness / larval cell width (c/w) has a

Figure 24 The T.cf. testaceitarsis has extremely thick connecters. They appear almost like chain links and sometimes still exist in the perimeter of the pupal brood

mean ratio of 0.364, thicker on average comparing cell connectors across the board. Larval brood is prolate spheroid shaped, and Pupal cells are Ovaloid.

A branching network of connectors is unique to *T. clypearis*, and these were observed in prolate spheroid larval cells. The branching network somehow 'dissolves' with the excess material probably utilized into the ovaloid pupal cells as the larva morphs. Otherwise, the callows will have a tough time emerging.

In some scenarios, the connectors are in filament form, especially in the new brood, as though in prep for more cells to be attached. This example image in Mindanao of a *Tetragonula* sp. with the species identity undetermined due to the uncertainty of the maturity of the nest structure.

Figure 25 connectors are in filament form

Figure 26 A branching network of connectors is unique to T. clypearis.

VIII. Mixed Oblate & Prolate Spheroids + Ovaloid shapes

In this particular type of nest, the brood cells appear in three stages, new brood cells are oblate spheroid, the intermediate stage is possibly the change to the pupal stage, and finally, the mature stage forms ovaloid type cells.

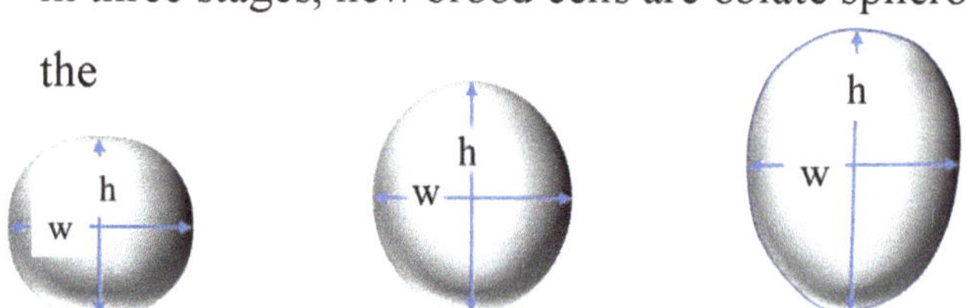

New Brood h/w range = 0.925 - 0.982	Pre-Pupal stage h/w range = 1.029 - 1.25	Pupal stage h/w range = 1.304 - 1.373

Figure 28 Average range of h/w over the Two nests were found.

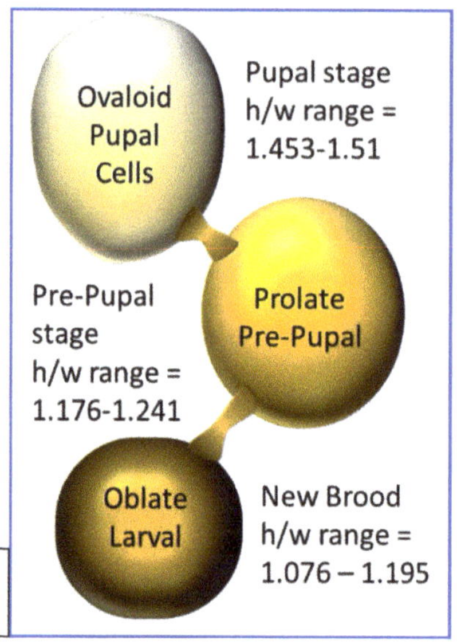

Figure 27 Mixed Brood-Cells

The *Tetragonula* sp. nests are currently undetermined species, and two nests were found of this nature in Wallacea. I captured the images personally in Julita, Leyte, Visayas, and Bukidnon Mindanao, where Gogo Manug contributed.

IX. Mixed Prolate Spheroids + Ovaloid & Cylindroid shapes

One more found in East Kalimantan near Samarinda with a combination of Ovaloid/Cylindroid (where the cell sidewalls vary from elliptically ovaloid to cylindrically parallel) Pupal, Ovaloid Pre-Pupal and Prolate Spheroid Larval cells. It is not a photographic depth issue; the oblique view shows distinct shapes and size ratio differences.

Figure 29 Mixed Prolate Spheroids + Ovaloid & Cylindroid shapes

Connectors: The connectors appear proportionately thicker in each different stage. At the pipal brood layer, the fringes have connectors where the c/w ratio = 0.544, about half the average pupal cell width. The pre-pupal c/w ratio range = 0.245 - 0.33, about a quarter to one-third of the average pre-pupal cell width. In the new larval brood, the average c/w = 0.26 max. about a quarter of the average larva cell width.

Chapter 4

4.0. Types of Brood Formation.

One slowly sees a pattern evolve from the relatively random and apparently chaotic mass. Observing and studying stingless bee brood, I try to unravel the clumps and delve into each component and peel its layer of form and arrangement.

4.1 Random clustered clumps *T.* cf *reepeni*

The general formation of brood cells in clumps clustered randomly. Some like *T.* cf *reepeni* appear gravelly, looking like gravel rather than like smooth pebbles. The dark-coloured cells are larval (*Figure 30*). They may appear unrelated from clump to clump due to different generations being formed (*Figure 32*), looking like little separate piles of gravel.

Figure 32 little piles of gravel – undetermined species

Figure 30 clustered clumps pf T. cf reepeni appear gravelly

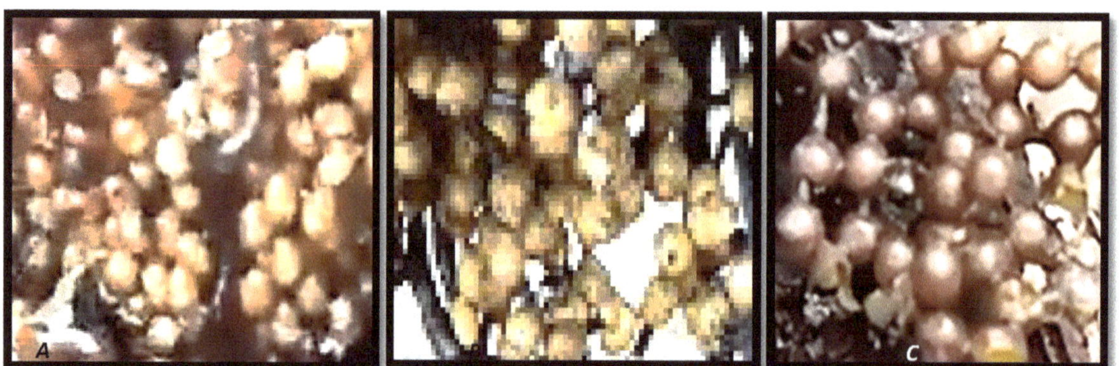

Figure 31 A. In Vietnam T. cf. iridipennis; B. Pupal brood of T. cf. iridipennis in Mindanao; C. Larval brood of T. cf. iridipennis in India

4.2. Loose and sparse - *T.* cf. *iridipennis*

Small population species may adopt a loose placement and allow cell expansion for pupal development. However, a smaller body size does not necessarily mean a smaller average population.

4.3 Whole Compact and Tight - *T. laeviceps*

They appear close together in an organized placement to have the greatest number of brood cells in the least amount of space. It looks like a dense formation in a tight limited space. One might expect this type of formation to occur in a bamboo cavity or a tight rock crevice. However, this formation is still prevalent in all *T. laeviceps*, even in a large wooden box cavity.

Figure 34 The organized placement looks like a dense formation in a tight space but is prevalent even in a large wooden box cavity. Images of T. cf. laeviceps in A. Panay; B. Sulawesi; C. Mindanao

Figure 33 Brood of T. cf. laeviceps in A. Thailand; B. West Papua; C. N. Borneo

4.4 Uniformly spread & dense - *T. fuscobalteata and T. clypearis*

Smaller species like *T. fuscobalteata* and *T. clypearis* are more uniformly spread than sporadic. They are minute but in large numbers.

Figure 35 Brood of T. clypearis in A. Lombok, Indonesia; B. in Negros, Visayas, Philippines; C. Cape York, Australia

Figure 36 Brood of T. fuscobalteata in A. Panay, Visayas, Philippines; B. in W. Kalimantan, Indonesia; C.in E. Kalimantan, Indonesia

4.5 Scattered & Sporadic - *Tetragonula ruficornis*

These may occur in young colonies where the population has not peaked. It would be difficult to determine the species from these broods.

Figure 37 Possibly young colonies and the population is low. Even the connectors are merely unattached filaments

Tetragonula ruficornis is generally the lightest coloured when directly examining the four primary types of the "iridipennis" species group in the Indian Subcontinent (Rasmussen, 2013). The bee closeups I see in some Vietnam videos appear very similar to the specimen images of *T. ruficornis* in the said paper.

4.6 Graduated Spacing - *T. mellipes*

So far, this type of graduated spacing of cells appears only in *T. mellipes,* where *the* larval brood is well spaced out but gradually gets denser as the stages develop into pre-pupal and tighter un the pupal stages. Its larval stages generally appear in the middle of the brood, and the pre-pupal radiate out and then the pupal brood are at the fringes of the whole big lump of brood cells.

Figure 38 graduated spacing of cells in T. mellipes

I was following one of Dr. Tobias Smith's previous lectures, and there was a mention that there could be two types of *T. mellipes.* Another source said they are waiting for results of DNA analysis that possibly three types of *T. mellipes exist.* I received this brood

Figure 39 A variation in the graduated brood of T. mellipes

cells image from Merauke near the West Papua/PNG border. So, it is left to the results of that DNA analysis to be seen whether there truly are three types of *T. mellipes* as the brood form variations show.

4.7 Zoned or layered stages *T.* cf. *Testaceitarsis*

The young brood is conspicuously zoned or separated from the mature brood in many instances. This obvious segregation can be seen in colouration, cell size, or colour and size. It is also seen in some species where the segregation pre-pupal stage is also segregated.

4.7.1. Zones: These can be in patches, clump, or terraced cluster levels. We can presume that the zoning characteristics will depend on the whole cavity size for the brood chamber. The zones generally are clearly defined visually

4.7.2. Layers: These are marked by connector thickness at the fringe of each layer. Unlike layered combs, this layered advancing front seems to be the Pupal batch at the top. The different stages of cell development are on different layers, and presumably, the layers move up as each batch of callow emerges. One wonders how a new batch of young brood begins at the bottom of the pile.

Figure 40 A. T. cf. Testaceitarsis with thick connectors at the pupal brood layer; B. T. cf. sapiens with cream pupal cells segregated from dark-coloured larval brood C. Terraced layers between the new and mature brood.

4.8 Cascading Brood - *T.* cf. *malaipanae*

Figure 41 Example of a cascading brood - T. cf. malaipanae

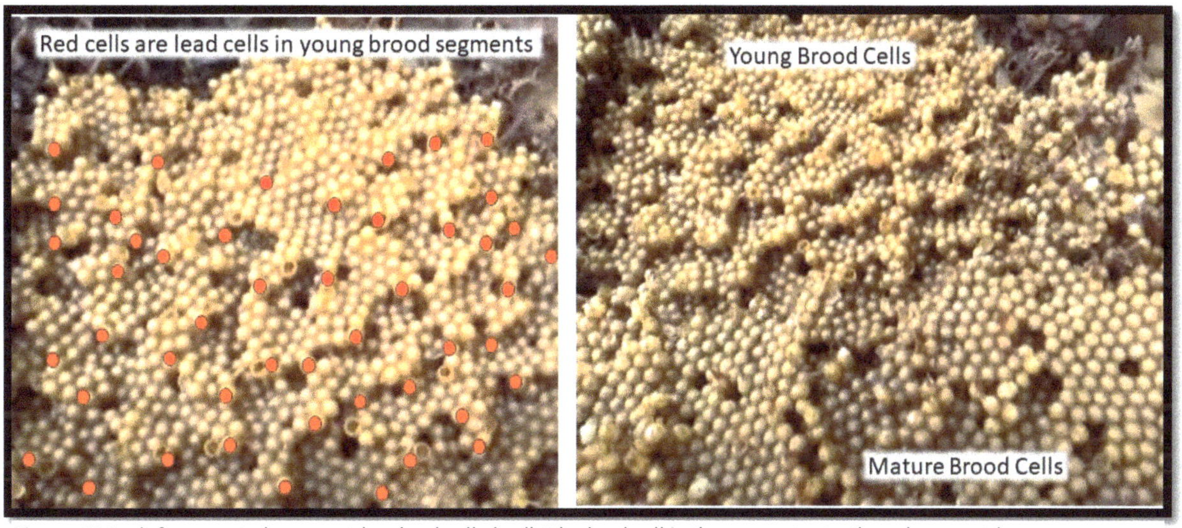

Figure 42 Each fragmented segment has lead cells (red). The lead cell is the connector to the subsequent lower segment.

This type, I would call a cascading brood - I suspect it as *T.* cf. *malaipanae* in Vietnam.

4.8.1 Cascading Brood Cells - Feature Peculiarity

1. Each fragmented segment has lead cells (red in *Figure 42*). The lead cell is the connector to the subsequent lower segment.
2. Unlike terraced layers, where each flat segment is laid on top of the other in a receding upward sequence, the terraced segments are connected by struts and pillars.
3. The cascade brood may look like a perforated brood form but missing the cell ring formation at an oblique angle.
4. The cross-sectional view shows irregular uplifting of segments, so it does not lead to any spiral form.
5. The young brood and the mature brood (semi-comb wafers) are separated in batches.
6. The building blocks appear like phalanxes but progress sideward with the lead cells.

7. Young brood cells are Cylindroid shaped with an h/w range 1.23 - 1.25; Mature brood cells are elongated Cylindroid shaped with h/w range 1.73 -1.83. The mature cells appear squeezed and elongated vastly compared to the young brood.

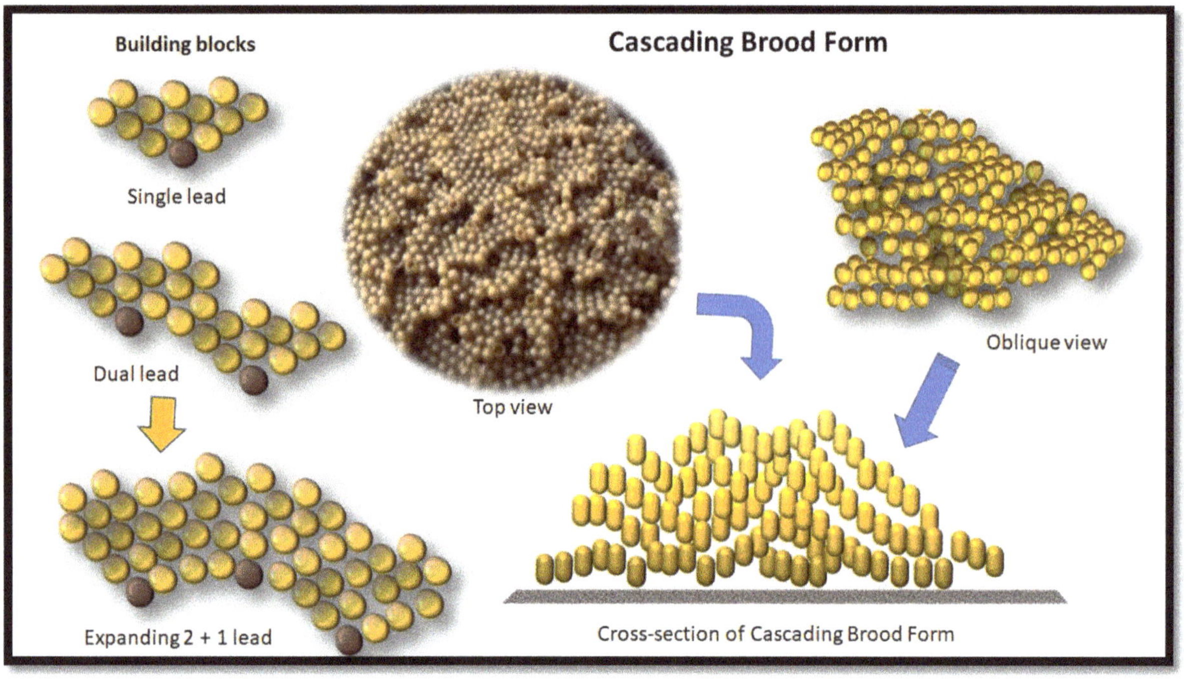

Figure 43 Cascading Brood Cells – Building Blocks and Orthographic projection

4.9 Chain-link Brood form - *T.* aff. *hirashimai*.

Looking at the diversity of *Tetragonula* brood, I am not sure what a clustered, semi-clustered or semi-comb brood is, especially when I look closely at this *T.* aff. *hirashimai.* I have yet to

Figure 44 Top: young and mature brood cells of T. aff. hirashimai with interspersed queen cells. Bpttpm: Tetragonula aff. hirashimai - the middle image is a queen and on the right is a gyne or a large callow.

describe this next odd-looking brood, but its size estimated bigger than most *Tetragonula* seems to lead me to *T.* aff. *hirashimai*.

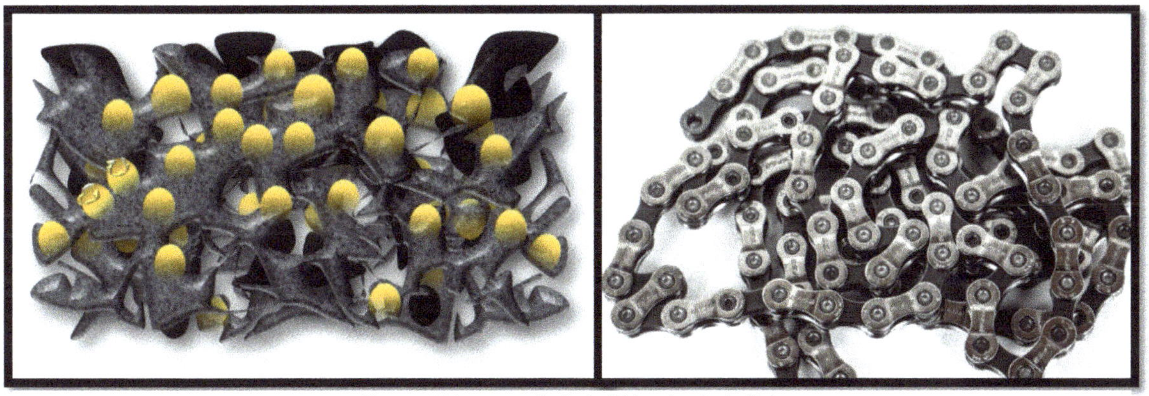

Figure 46 I am reminded of bicycle chain links

I am reminded of bicycle chain links. The links are haphazardly random, and the cells appear at junctures and may either be a queen cell (h/w ~ 1.17) or a normal larval cell (h/w ~ 1.268). The queen cells are slightly bigger and stouter, and all cells are Ovaloid shaped. Of the images I have, the brood size is nominal, and they appear in levels 3 or 4 at most. These are newly cultured in box hives and may not have fully developed yet.

Figure 45 This is also bicolorous; I can only assume it is either T. drescheri or T. sarawakensis.

4.10 Miscellaneous brood forms.

4.10.1 Ovaloid clustered Clumps with Ellipsoid Queen cells – *T. iridipennis?.*

Figure 47 tightly packed clump of brood cells with a conspicuous queen cell.

I received is this clump of brood cells with an obvious queen cell among the peculiar brood forms. It is from Arvie Aguilar Misolas Daraga, Bicol Region, Philippines. The Ellipsoidal queen cell h/w ratio = 1.44 and the pupal brood cell h/w range = 1.212 - 1.323 while the larval cell h/w range = 1.225 - 1.36. All pupal and larval cells are Ovaloid shapes. This form is uncommon as far as the Bicol region is concerned. I would compare this with a similar form in India, namely *T.* cf. *iridipennis*. The more common *T. iridipennis* in India is not as tightly packed as this, but the brood cell shapes coupled with queen cell shape is the same as this form in Bicol Philippines.

Figure 48 T. iridipennis in Karnataka, India

The image from Beekeepers of Karnataka is by courtesy of Dr. Bhagarva Hukunda shows the queen cells average h/w ratio = 1.378 Ellipsoid and the pupal brood cell h/w range = 1.231 - 1.38 while the larval cell average h/w ratio at 1.2 a similar configuration, but the h/w ratios are not close enough to constitute a category of its own. The h/w ratio values do not indicate the respective size in relation to each other. I decided to look into the queen cell width to pupal cell width ratio and the queen cell width to larval cell width ratio, and the pupal cell width to larval cell width ratio. See table below.

Table 1 Relation among Queen cells with Pupal cell and larval cell widths (W)

Image source	Q-cell W/Larva W	Q-cell W/Pupa W	Pupa W/Larva W
Bicol, Philippines.	2.24	1.65	1.07
Karnataka, India	2.275	1.453	1.15

This tabulation data shows a closer similarity between the brood cell configuration of the two mage sources. Comparing the width ratios gives a perspective of the sizes of each type of cell in association with each other.

4.10.2 Small Scattered clumps - *Tetragonula* aff. *basimaculata*

These images are from 2019 – *Tetragonula* aff. *basimaculata* (Bingham, 1903) {Taxonomic Serial No.: 1129864 - ITIS Report}. I received what I believe could be *T. basimaculata* from Vietnam.

I do not have microscopic images of all those bees. I need to collaborate with someone in Vietnam, and I only communicate with Google translate with my sources, and it is not an easy task

Figure 50 Tetragonula basimaculata (Bingham, 1903) Taxonomic Serial No.: 1129864 - ITIS Report

The *T.* aff. *basimaculata* brood cell h/w ratio ranges from 1.184 to 1.27 Ovaloid shape.

Figure 49 I received images of what I believe could be T. basimaculata from Vietnam.

Chapter 5

5.0 Semi-combs

These are in small groups of 15 to 25 cells bunched like small combs. They are then placed in layers, and some are terraced, and some may appear curved rather than all flat. The examples of these are *T. pagdeni*. *T. hockingsi* and *T. davenporti*

Figure 51 Semi-combs A) T. drescheri, Sunda, Indonesia; B) T. hockingsi, Qld., Australia; C) T. pagdeni, Thailand, SEA

These are my preliminary findings, studying many of what has been termed semi-comb brood.

The workers work on certain building blocks, and the key is to determine these building blocks. They start with the typical hexagonal standard.

The ring of cells may be a five cells ring, six cells ring or seven cells ring or a combination of these. The bigger rings can accommodate larger queen cells when the need arises. The random configuration results in a perforated brood combining the various rings. There could be loose perforations or tight perforations. A combination of the two types of perforation then results in a perforated brood that I think resembles the *T. mellipes*' Graduated Brood', just something to ponder about.

I speculate this phenomenon avoids the fusion of adjacent cells and controls queen cell occurrence. e.g., see the image below.

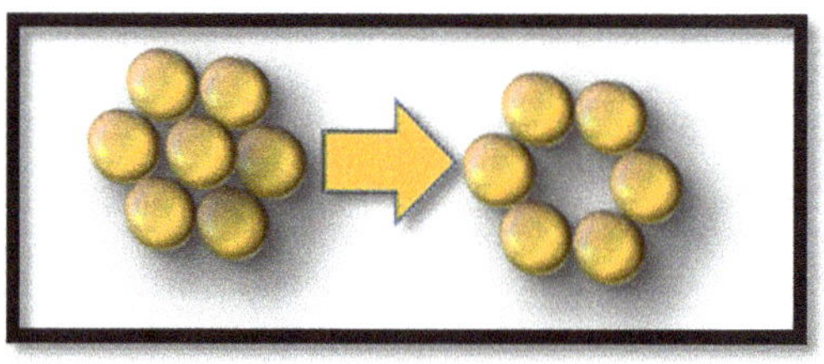

Figure 52 typical hexagonal standard.

5.1. Perforated Brood - *T. pagdeni.*

When loose perforated groups of Ovaloid cells lay one on top of the other, it results in a layered perforated form as in *T. pagdeni* (Figure 53 & Figure 54).

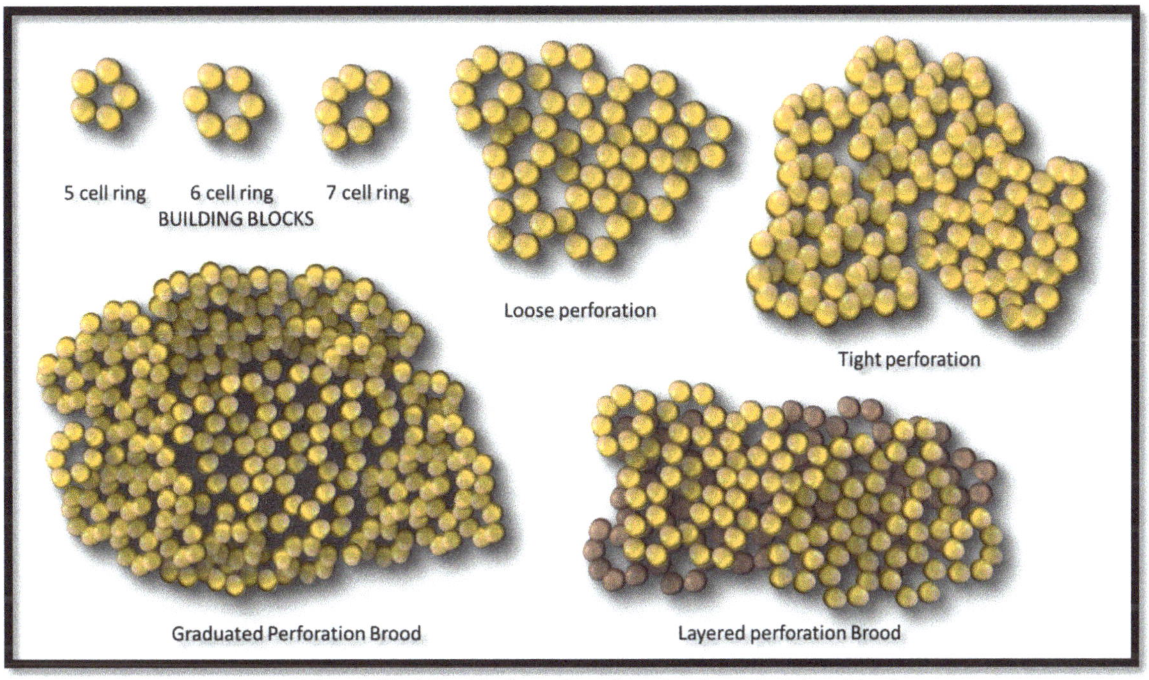

Figure 54 set of building blocks and combination forms in Spheroid cells

Figure 53 Semi-comb brood in Vietnam

5.2. Ordinal Brood - *T. drescheri*

Another set of building blocks I relate to is the military formations. This kind, I see in broods that remind me of the historical Macedonian Battle Formation.

I call them Ordinal because I see some order in the seemingly chaotic formation. This form is quite obvious in groups of spheroid cells like *T. drescheri.* Although interspersed, each semi-comb is almost regimented, and the whole placement appears like an orderly brood (*Figure 55*).

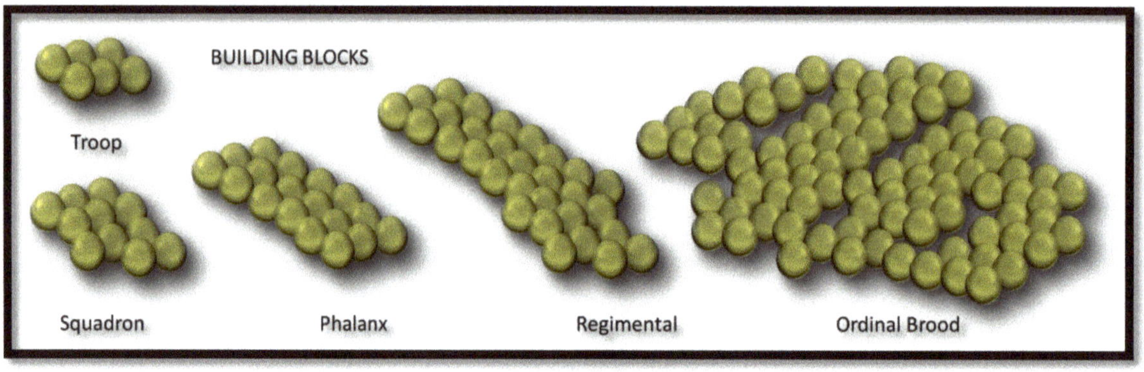

Figure 55 set of building blocks and combination forms in Ovaloid cells

5.3 Layered Wafer Brood - *T. biroi and T. carbonaria.*

The other set of building blocks is of the group of Cylindroid or Ellipsoid cells. This type occurs in the early or young brood of *T. biroi and T. carbonaria.*

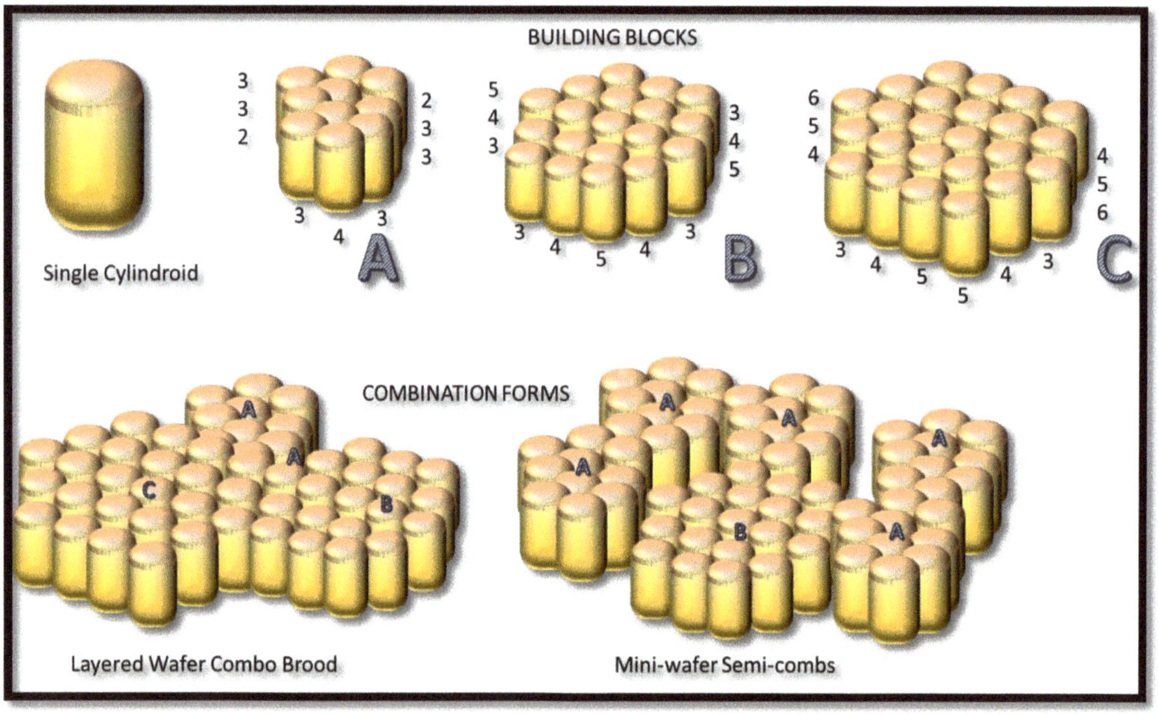

Figure 56 set of building blocks and combination forms in Cylindroid cells

The building blocks here are hexagonal; A. is 3/4/3 x 2/3/3/2 Hex; B. is 3/4/5/4/3 x 3/4/5/4/3, and C.is 3/4/5/5/4/3/ x 4/5/6/5/4. (C. looks like a silicon microchip wafer). Many combinations of these building blocks may vary depending on the available cavity shape. 2 types of wafer combo, the Layered wafer and the Mini wafer semi-combs types. These combination forms occur in *T. hockingsi* and *T. davenporti* and early stages of full comb broods. Other variations are in different positions in a hive cavity.

5.4. Segmented semi-comb Brood – *T. hockingsi*

The pyramidal formation is in the centre of a hive box, while a terraced formation is at the corners. The stacked type appears for *T.* aff. *sarawakensis* and others that are in tight spaces or circular cavities.

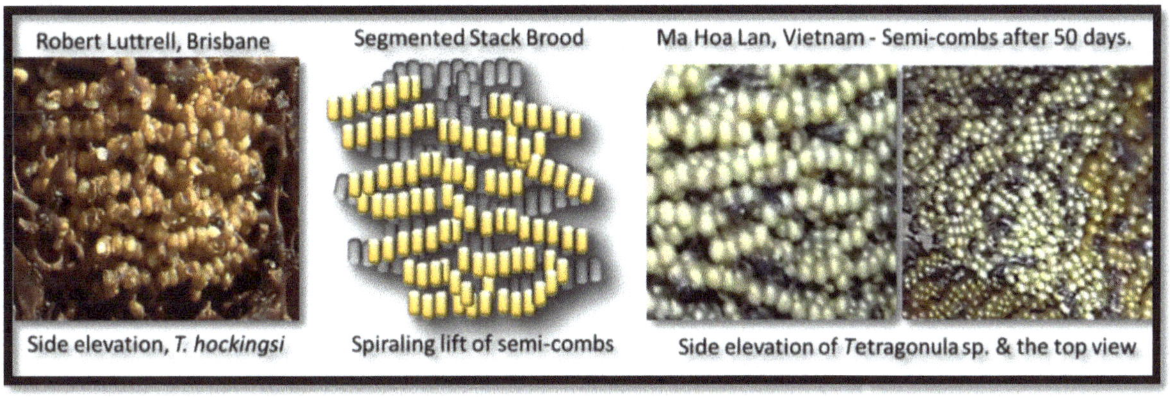

Figure 57Side view of segmented lift in Tetragonula hockingsi and sp.

On the right is an image of a perforated semi-comb similar to *T.* cf. *pagdeni* or some other *Tetragonula* sp. or complex of species in Vietnam; both side elevations show an apparent spiralling lift of the semi-combs. We call it a Segmented Stack brood.

5.5 Terraced semi-combs – *Tetragonula* sp.

Reminds me of rice terraces; these undetermined species so far only observed in Vietnam.

Figure 58 Reminds me of rice terraces these undetermined species, so far only observed in Vietnam.

5.6 Mixed Semi-combs Ω clusters - *T. drescheri*

For example, *T. drescheri* brood cells are of two types, cluster types in the young colony. In this type, the brood cells are arranged in one group. Semi comb, i.e., comb-like form, but the places is irregular (Salmah 2017). This phenomenon may also occur in *T. sarawakensis* or *T. pagdeniformis*.

Figure 59 Large population brood cells appear in two types, cluster types in one group. semi comb or comb-like form in another and the places arrangement is irregular

A situation may also occur where a particular brood formation species takes over from another formation altogether. Examples are like *T. laeviceps* taking over *T.* nr. *biroi* (seem in Java) or *T. hockingsi* taking over *T. carbonaria* in Australia. It may not commonly occur in their natural habitat, but it is common in Meliponaries where forage vegetation is insufficient, creating territorial competition and unbalanced resource partitioning. Besides forage competition, it may also be caused by weakening one (e.g., initial overwhelming by pests or parasites or for whatever other reason) and then taken over by another stronger colony. See Figure 61 of *T. hockingsi* taking over *T. carbonaria* in Queensland, Australia ("Splitting hives in Brisbane today I found a *T. carbonaria* hive that looks like a *T. hockingsi* has nearly overtaken it", Mathew Middleton).

Figure 61 "Splitting hives in Brisbane today I found a TC hive that looks like a T. hockingsi has nearly overtaken it", Mathew Middleton.

Figure 60 *Comparing T. drescheri in Purworejo, Java*

5.7 Mixed Spheroids Clusters morph into Semi-comb Ovaloid – *T. cf. minor*

It has been observed in an example of *T.* cf. *minor* (see *Figure 62*) a Perforated cluster brood morphs into an Ordinal arrangement within two months. The larval brood cells morphed into cocoons after 60 days. The pre-pupal brood (Avg. h/w = 1.218 ± 5%) above and the (Avg. h/w = 1.3165 ± 2%) pupal brood.

Figure 63 Ong Dú Viet Nam *– Da Lat, Lâm Đồng – Left: Larval brood cells morphed to pre-pupal. Middle: pre-pupal brood (Avg. h/w = 1.218 ± 5%) and Right: (Avg. h/w = 1.3165 ± 2%) pupal brood after 60 days.*

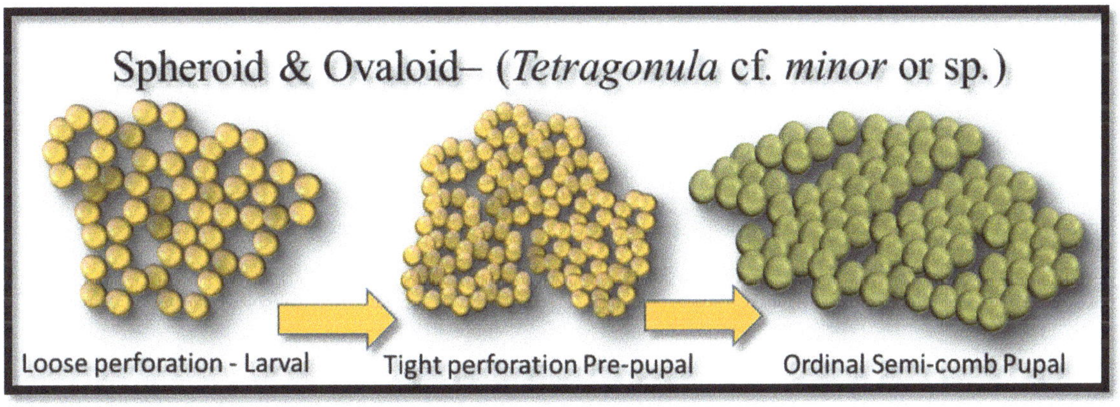

Figure 62 Spheroid to Ovaloid morph in T. cf. minor.

Truong Nguyen shared a post to the group: *Hội ong dú toàn quốc*[3]; https://www.facebook.com/groups/2134975603386858/ I can only suspect *T.* cf. *minor* because of the proportion of the bee size with the early cell size and the different stages h/w ratios corresponding to previous data that I have. It exhibits similarities with *T. drescheri* in brood arrangement in the different stages, but *T. drescheri* brood cells are oblate spheroid throughout. *Tetragonula minor* pupal brood cells, *T. ruficornis* and also *T. melanocepahala* form reminds me of a corn cob. The images here is contributed by Ahmad sapruddin in Banjarbaru, S. Kalimantan vide https://www.youtube.com/watch?v=lOprPjWIihI.

Figure 64 Left: A corn cob; middle: T. melanocepahala; right: T. ruficornis

[3] Dr. Pham's note: I can explain a group of Vietnamese words: " Ong dú " means stingless bees. The "Hội ong dú toàn quốc " that you mentioned, is just some meliponicultura only on facebook, it is not an official association of Vietnam. So, you should remain the "Hội ong dú toàn quốc " without translation in English.

Chapter 6

6.0 Full combs layered, spiralled or terraced

Relevant regions are as follows:

1. in **Sunda** are the likes of *T. drescheri* and one more that is often mistaken for *Tetragonula* but is *Sundatrigona moorei* (See Figure 65 Left).
2. in Nagaland, **India** is a *Tetragonula cf. kyrdemkulaiensis* (See Figure 66 Left) and in Kerala S. India possibly *T.* cf. *laeviceps* complex (See *Figure 72* Bottom Left);
3. in Mainland SEA is *Tetragonula laeviceps* complex (also possibly complex of species: pagdeni and pagdeniformis) in Thailand, also in Cambodia a *Tetragonula aff. sarawakensis* very similar to the one in Nagaland (See Figure 66 Right);
4. in **Sahul** is *T. carbonaria* (See Figure 67)
5. in **Wallacea** is *T. biroi* or *T.* nr. *biroi* (See Figure 76);

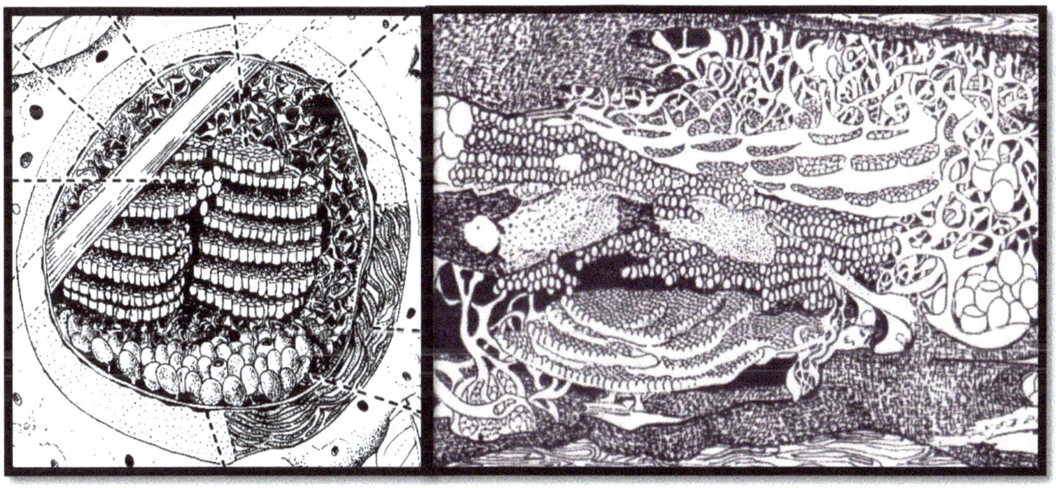

Figure 65 Left: Nest architecture of Sundatrigona moorei within arboreal ants' nest, traced from Sakagami et al. 1989; Right: Nest architecture of T. drescheri in a box, traced from Salmah & Yasmi, 2017

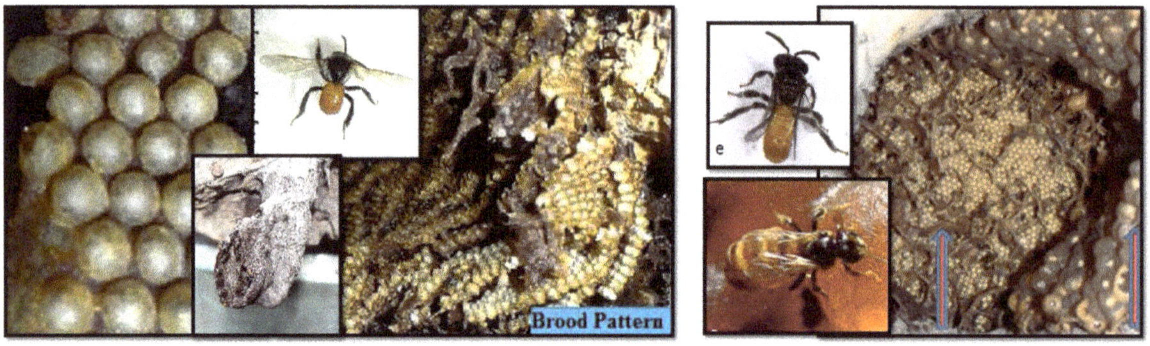

Figure 66 Left in Nagaland, India is a Tetragonula cf. kyrdemkulaiensis from Thungben Yanthan; Right: in Cambodia a Tetragonula aff. sarawakensis by Stephen Peterson.

Figure 67 Phases of Tetragonula carbonaria A. First stage of new brood; B. comb formation and terraced groups; C. fusing of the groups to a spiralled form; D. side elevation; E. concaved spiral; F. Bottom view. Images by Mathew Middleton

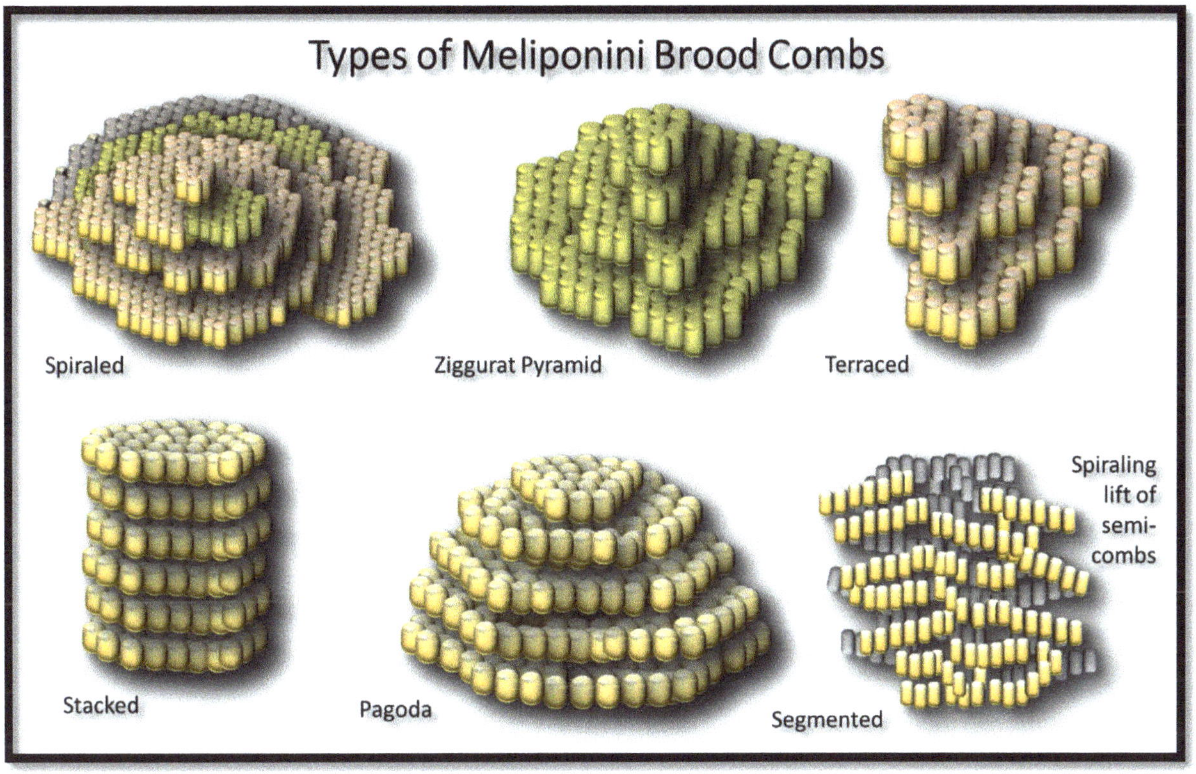

Figure 68 Types of Meliponine Brood Combs

6.1 Pyramidal Form - Stacked Pagoda & Stepped Ziggurat

Figure 69 from Bob Luttrell of his T. carbonaria in a cylindrical hive

The Pagoda type is a new classification I added when I received an image from Bob Luttrell of his *T. carbonaria* in a cylindrical hive. The spiral type does not occur commonly and is only 10% of *T. biroi* and *T. carbonaria* (put together as far as my image bank is concerned). I have not seen any spiralling in the full comb brood of Nagaland. When I drew the spiral type comb, it was three times as hard, with amendments and redraws, and I only managed one satisfactorily. If drawing can be related to building a spiral brood arrangement, I can only imagine how difficult it must be for the bees. Yet, they do it, but why? If 90% can make do with regular layering of their cells, no matter whether step-pyramidal, terraced or pagoda stacking. There does not seem to be a necessity to spiral up.

6.2 Spiral Brood – Sinistral or Dextral

I tried considering the Coriolis[4] effect, but *T. biroi* is in the northern hemisphere, yet it has both sinistral and dextral spirals. Similarly, *T. carbonaria* is in the southern hemisphere, and it also has both sinistral and dextral spirals. Then for *T.* aff *sarawakensis* and *T. kyrdemkulaiensis*, Viraktamath and Rojeet sp. n., where both are in the northern hemisphere and no spiralling has been observed or reported. So, I can deduce that the Coriolis force does not come into play in this instance.

I determined at the start that I had only 10% of spiralling brood comb in my image bank for full combs, so I solicited images of *Tetragonula* spiral brood in social media interest groups. I found very intriguing results. Initially, I kept only a few *T. carbonaria* images because classifying the full layered brood combs were not an issue. But, of late, I received quite a number, and for *T. carbonaria* in my image bank is now 34 in total... out of which twelve are spiralled (eight sinistral & four dextral), three pagoda stack types, four segmented brood, two ziggurat pyramid type, nine terraced wafer layers. That means almost 30% is spiralling (data collection is ongoing).

For *T. biroi* and *T.* nr. *biroi*, my image bank is much better now. The Philippines is now 75 in total with thirteen spiralled (five sinistral & eight dextral), 41 segmented...three twin peak ziggurats...18 terraced wafer types. This data makes the spiralling at more than 17%. From Indonesia, I now have thirteen in total. One pagoda stack. There were seven segmented mini wafer broods (one from West Papua)... five terraced wafer layers, and none were spiralled.

Figure 70 sinistral (left from Dan Smaoles, Sydney) and dextral (right from Mathew Middleton, in Markham Valley) spiral in T. carbonaria

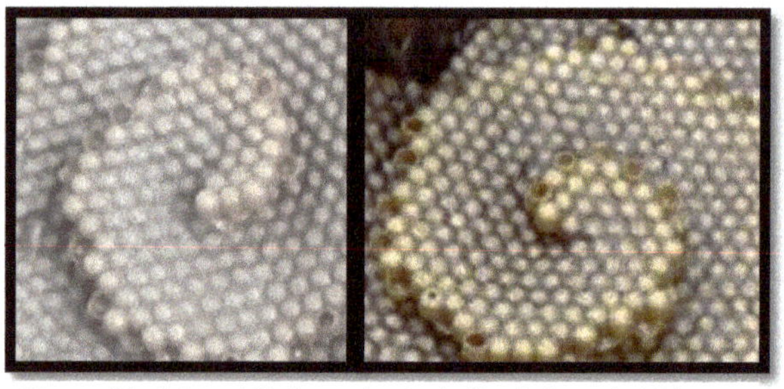

Figure 71 sinistral & dextral spirals (both from Lui Morota, Alaminos Farm, Laguna) in T. nr. biroi. I was further astonished when I saw this one

[4] Coriolis force - an effect whereby a mass moving in a rotating system experiences a force (the Coriolis force) acting perpendicular to the direction of motion and to the axis of rotation. On the earth, the effect tends to deflect moving objects to the right in the northern hemisphere and to the left in the southern and is important in the formation of cyclonic weather systems.

Now, what have all these got to do with the sinistral or dextral spiral brood of *Tetragonula*? I don't know yet whether there is any significance[5].

The bee builds the brood in 4 quarters in the larger cavity boxes. In this particular case, different types of advancing peaks in the same hive. The possibility of subspecies for different types of broods (as some of my comrades in beekeeping like to speculate) just went out the window.

6.3. Curved Brood – Convex or Concave

So, we come to the variations in the types of meliponine brood combs. A convex top was found in a box hive in Bohol, Philippines. However, I presume this is an anomaly because only one in a total of 118 *Tetragonula* comb brood images from India, Indochina, Wallacea and Sahul. I say this is an anomaly as it could indicate something amiss. Can excess heat cause a convex curl up of the top layer?

Figure 72 Top: A convex top brood in a box hive in Bohol; Bottom Left: convex brood in a bamboo hive in Kerala, India; Bottom right: in a circular cavity in Medziphema, Nagaland.

Other convex forms appear moulded by the cavity shape, as seen in the cylindrical cavities in India. (See Figure 72 Bottom Left & Bottom Right)

[5] Editor's note: a spiral saves material (the pillars) and reduces 'vulnerable space' between combs- my observation DWR

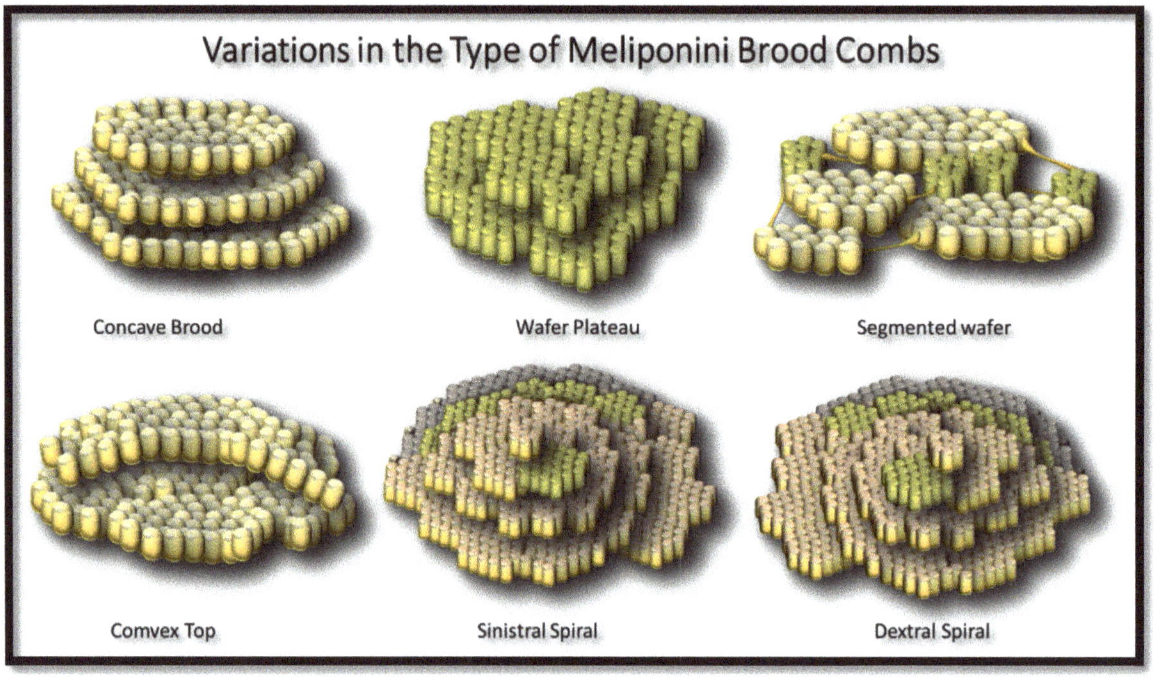

Figure 73 variations in the Type of Meliponine Brood Combs

6.4 Wafer Brood – Segmented or Plateau

Figure 74 Examples of wafer plateau brood in the Philippines

The wafer plateau may have only 2 or 3 layers of a wide expanse and may be influenced by the height of the 'supers. The segmented wafer type is found in Vietnam, Wallacea, to Australia regions. They may be flat or curled, and the connectors (besides pillars) keep them in place relative to each level. Beekeepers told me in Sulawesi, and this is the early stages of full combs in their *T.* nr. *biroi* Sulawesi.

Figure 77 variations of early-stage T. nr. biroi segmented brood in the Philippines

Figure 76 Feral colony of Tetragonula nr. biroi in Romblon, Philippines. Images by Joe Asis

The segmentation is also similar to the early stages of *T. carbonaria*

Figure 75 On the left, Lockyer, QLD and on the right, Sydney, NSW.

For *T. carbonaria*, it does not build up this way. I see a sequence below with Matthew Middleton· Lockyer, QLD., and Dan Smailes, Sydney, Australia. (See Figure 67 Phases of *Tetragonula* carbonaria A. First stage of new brood; B. comb formation and terraced groups; C. fusing of the groups to a spiralled form; D. side elevation; E. concaved spiral; F. Bottom view. Images by Mathew Middleton) The brood starts clustered (I gathered from Mathew Middleton) then into segmented layers. From the segments, they can go either clockwise or anti-clockwise (ambidextrous, I guess). This illustration is my impression of the segmented lift to a spiral sequence. Struts and connectors hold the positions with only a few initial pillars. From this segmentation develops the dextral spiral.

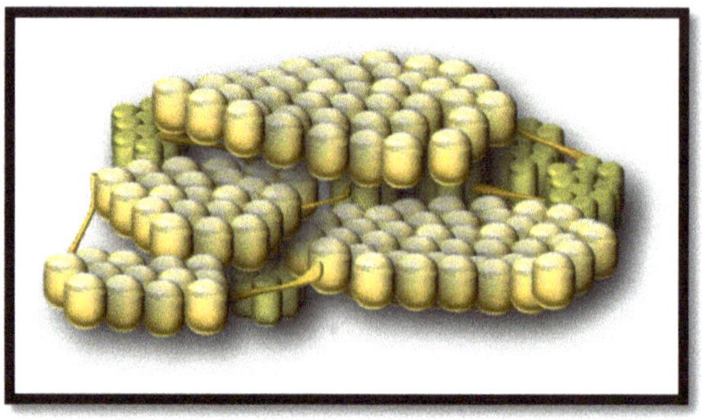

Figure 78 Segmented lift to a spiral sequence.

6.5. Morphing brood form

I had the opportunity to test this hypothesis on brood cell shapes on a brood form query from Sam Perry of Brisbane. Figure 79 shows Sam Perry's brood form query, whether *T. carbonaria* or *T. hockingsi*. I enlarged the image to get the brood cell shape and sized the h/w ratio. The brood cell shapes are all Cylindroid h/w = 1.152 ± 3%. I compared these data with *T. carbonaria*.

Figure 79 Brood form query on species identification

Based on brood form development, as reported by Matthew Middleton, he observed the growth of *T. carbonaria* brood from scratch. It starts as a randomly clustered brood and then forms semi-combs or mini-wafers and later morphs into one of three, i.e., concave layered, dextral or sinistral spiral. The brood cell shapes are Ovaloid to Cylindroid depending on whether they are compacted within the layers or are at the layer perimeter. The brood cell h/w range = 1.326- 1.378.

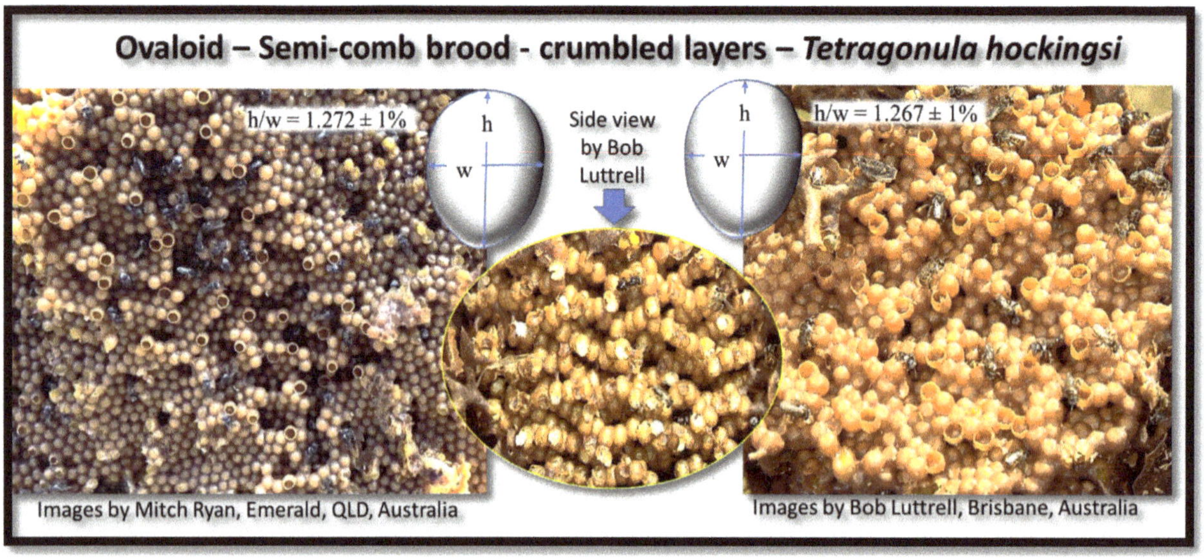

Figure 80 T. hockingsi. semi-comb brood forms

I then compared *T. hockingsi* semi-comb brood forms, which are in crumbled layers. The mini combs are not flat. See samples above from Bob Luttrell and Mitch Ryan. The *T. hockingsi* brood cell shapes are Ovaloid, and the cell h/w range = 1.267 - 1.272 in both samples.

Next, *Figure 81*, a brood form that I photographed when visiting Peter Davenport in Elenora. Here the semi-combs are flat, and all the brood cell shapes are Cylindroid throughout. The brood cell h/w = 1.187 ± 3%. So now, looking at this and matching it to Sam Perry's photo *Figure 79*, they have geometric shape similarities and closely resemble the h/w ratio. Hence, in my opinion, Sam Perry's brood is *T.* cf. *davenporti*. We will ask him to develop it for a couple of months and then follow up to see what form it morphs into.

Figure 81 Images by AHJ @ Peter Davenport's personal hive, Qld., Australia

6.6 Varied Combo Brood – Quadrants or Triads

Figure 82 In T. biroi, the same thing happens with the one from Makassar, Sulawesi, the early segmented stag, although the different forms rarely appear in the same nest.

In *T. biroi*, the same thing happens, I think, as they appear in varied forms.

The one from Makassar, Sulawesi, is the early segmented stage. In Indonesia, the different forms rarely appear in the same nest.

However, in the Philippines, many different forms appear concurrently in the same nest.

Figure 83 On the left is the early segmented brood.

6.7. Subterranean Dwelling *Tetragonula*

6.7.0. I would like to acknowledge Dr. Viraktamath for his January 2020 paper https://www.researchgate.net/publication/350710892_Two_new_species_of_*Tetragonula*_Hymenoptera_Apidae_Meliponini_from_North-East_India_with_notes_on_their_nest_structure and images of their brood cells and also the *Lisotrigona chandrai* brood cells he sent me earlier. Although this paper has accurate morphometry, I still have to keep it consistent by using my PowerPoint sizing feature on those images. All the images of layered brood images in Nagaland that I suspected as *Tetragonula* aff. *sarawakensis* could very well be what the said paper establishes as the new species *Tetragonula kyrdemkulaiensis* Viraktamath and Rojeet sp. n.

6.7.1. This *T. kyrdemkulaiensis* was found in an underground nest at Kyrdemkulai, Ri-Bhoi District of Meghalaya, in January 2019. The nest was located at a depth of 90 cm in the soil with an entrance burrow (116 cm in length) running horizontally. See *Figure 84*.

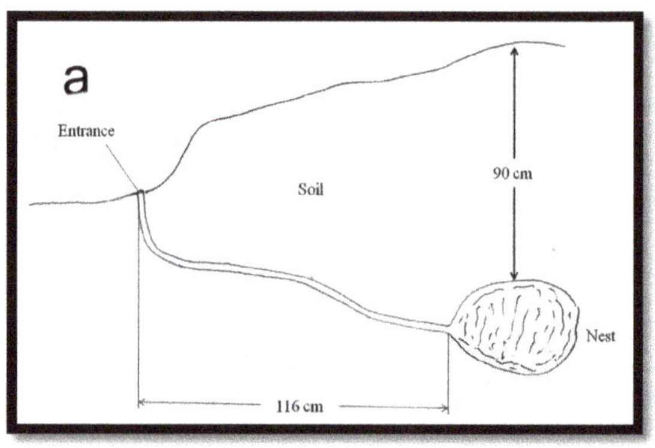

Figure 84 subterranean nest at Kyrdemkulai,

6.7.2. In South East Asia, there is the tight and organized brood form of *Tetragonula sirindhornae.* Sometimes found subterranean even in car park perimeter curbs. It has *an* Ellipsoid, Ovaloid & Spheroid combination. A unique cell combination where new cells are spheroids in dark brown. The larva develops as the egg hatches within, and the cell turns beige. The spheroids then develop into a pale ovaloid shape in the pre-pupal stage. Once the pupa matures, the cell develops a cream Ellipsoid shape. The connectors remain the same thickness throughout, although the c/w ratio gets lower as the cell width grows later.

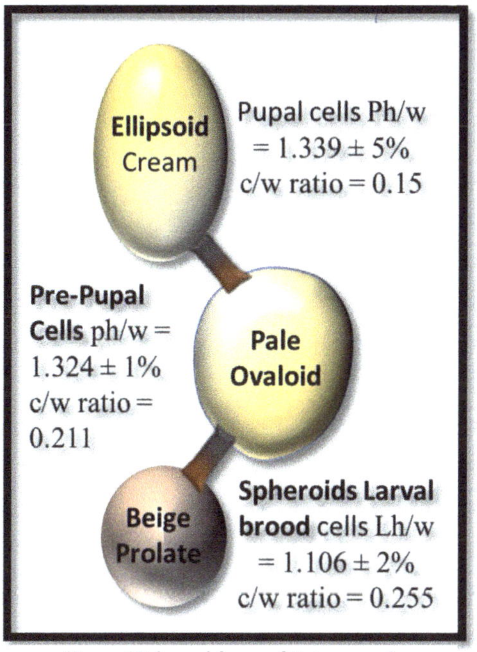

Figure 85 brood form of Tetragonula sirindhornae.

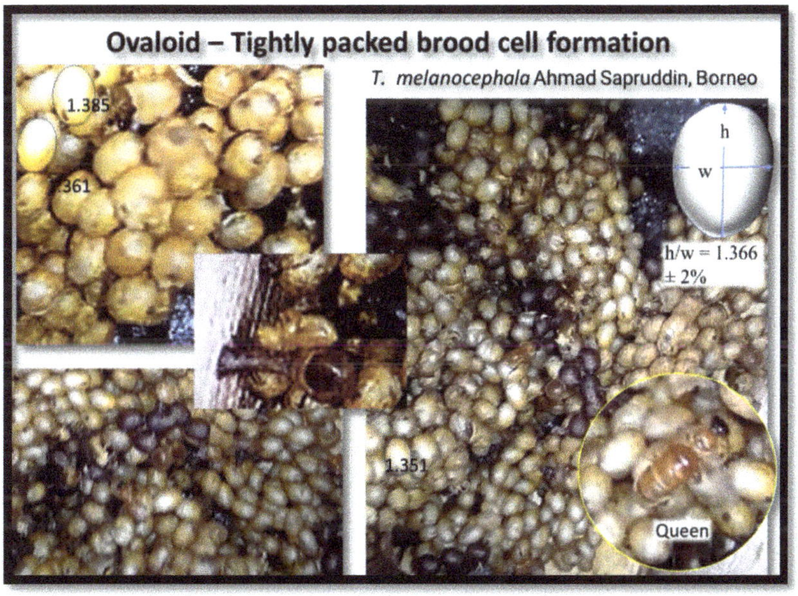

Figure 86 tightly packed brood formation of Tetragonula melanocephala

6.7.3. In Borneo, we have a tightly packed brood formation of *Tetragonula melanocephala,* and it is almost exclusively subterranean. I have not seen this species nesting above ground, i.e., rock outcrop or arboreal. In many instances, the nests were found with the nest entrances protruding upwards from the ground.

6.7.4. Another exclusively subterranean is *Tetragonula (Tetragonilla) atripes.* Usually found in the Sunda Region of Peninsula Malaysia and Sumatra. This species, too, has yet to be found nesting above ground. This Subgenus *Tetragonilla* is dealt with in the next chapter.

6.7.5. For *Tetragonula* aff. *sarawakensis*, the brood cells are mostly wafer layers, and I have not seen any spiralling, but they sometimes show they clustered new cells in the young brood.

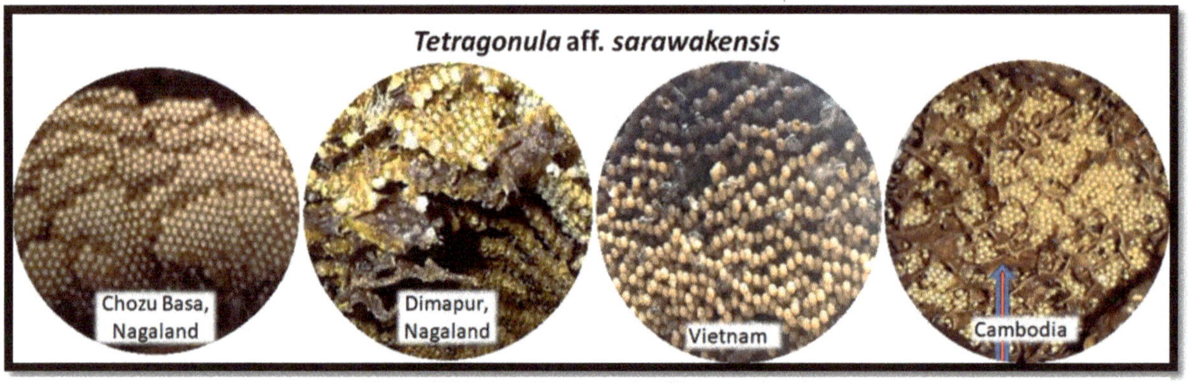

Figure 87 Various Tetragonula aff. sarawakensis

They are usually subterranean in Borneo and Cambodia, found in a termite's mound.

All said and done, and I have a nagging question. Why is it the underground dwelling *Tetragonula* have a red abdomen? I dub them "Ladies in Red".

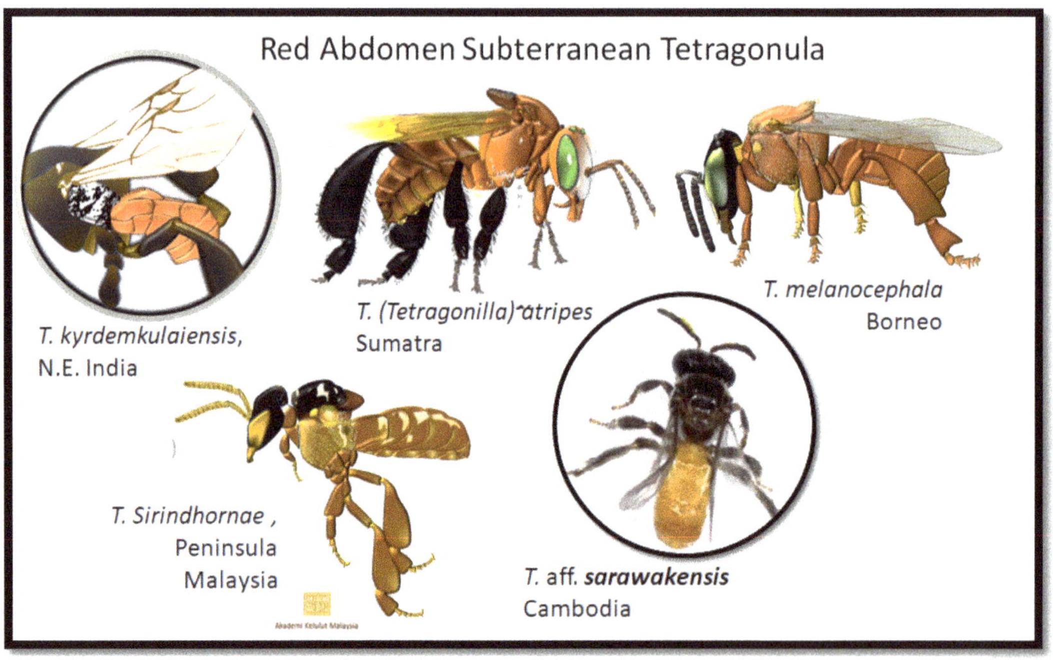

Figure 88 Subterranean dwelling Tetragonula have a red abdomen "Ladies in Red".

6.8. Regimented Pupal Brood

The cocoons are meticulously arranged, seemingly equally spaced in uniform rows and columns; I call it the Regimented Brood. The larval cells are randomly clustered with Ovaloid to Ellipsoidal shapes, but the cocoons are Cylindroid and so well arranged you could not have set them better yourself manually. I am not able to determine the species yet. (See *Figure 124*

The cocoons are equally spaced in uniform rows and columns; I call it the Regimented Brood - Tetragonula sp.) (See also Chapter 10, Item 10.8 Regimented Brood, page 75)

"The equal spacing comes from 4-piece stabilizers. The mass will not shrink when the cerumen of cells is removed. And it is the placement up against the wall itself- which has nothing to do with the pupal stage per se. The pupae had egg and larval cells in the same place and the same formation beforehand. It is attached to a wall and each cell to four others. Those outside this layer MUST be configured differently" (DWR).

Figure 89 Regimented brood placement at box side wall

Chapter 7

7.0. *Tetragonula sensu lato*

7.1. Subgenus *Tetragonilla*

7.1.1. *Tetragonula (Tetragonilla) atripes* in Selangor, Peninsula Malaysia.

1. Larval cell height to cell width ratio (L-h/w) = 1.326 ± 5% are Cylindroid but inclined to Ovaloid and are dark brown in colour, larval stage connectors to cell width ratio (c/w) = 0.22
2. The pre-pupal stage gets cream coloured.

Pre-pupal cell height to cell width ratio (p-h/w) = 1.235 ± 2%

3. The Pupal cells eventually turn reddish-brown,

possibly due to the pupa-callow pigmentation[6].

Pupal cell height to cell width ratio (P-h/w) =

1.484 ± 3%

4. The connectors appear fairly consistent (about equal in all stages but result in a lower value in Pupal cells) with a c/w ratio = 0.12

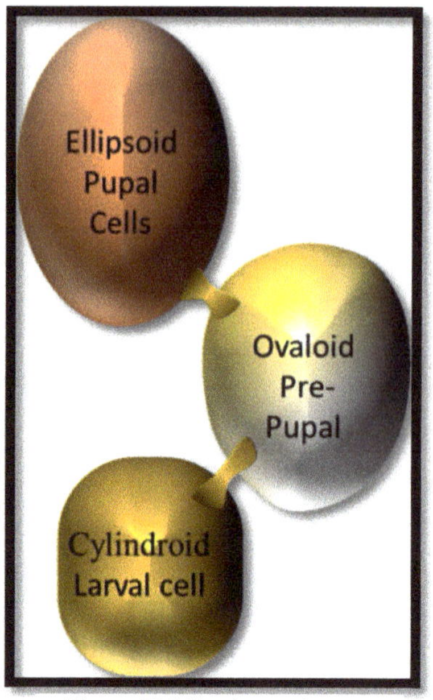

Figure 90 Mixed egg shapes - Ellipsoid, Ovaloid & Cylindroid

Figure 91 Tetragonula (Tetragonilla) atripes in Selangor, Peninsular Malaysia

[6] Editor's note: Why are they so red? resin? AHJ: the redness in the cerumen are possibly exudates the bees forage from Red Balau, a Rubroshorea https://www.lesserknowntimberspecies.com/species/balau-red

7.1.2 Brood cells of *Tetragonula (Tetragonilla) collina* in Vietnam

The farmers in Noong Het commune, Dien Bien district, Dien Bien province helped collect stingless bee *T. collina* colonies and several former student assistants who moved these colonies to the VNUA campus from Dien Bien province.

Table and images courtesy of Dr. Pham Hong Thai. Hanoi

Figure 92 Brood cells of Tetragonula (Tetragonilla) collina in Vietnam

Table 1: Worker cell dimensions of *Tetragonilla collina*

Parameters	Cell length (mm)	Cell width (mm)
Average	5.3 ± 0.34	4.28 ± 0.37
Min	4.4	3,9
Max	5.9	5
n	31	31

Tetragonula (Tetragonilla) collina in other regions

Figure 93 Tetragonula (Tetragonilla) collina in other regions

Figure 94 Tetragonula (Tetragonilla) collina in Perak, Peninsular Malaysia

Chapter 8

8.0. Other Genera

8.1. *Austroplebeia* brood structure

8.1.1. *Austroplebeia australis* larval brood image from Dean Haley, Brisbane, shows the cells in small random groups of 12 to 16. They have clustered clumps, not layered semi-combs. The larval cells h/w = 1.177 ± 1% and the pupal cells h/w = 1.243 ± 1% with c/w = 0.168.

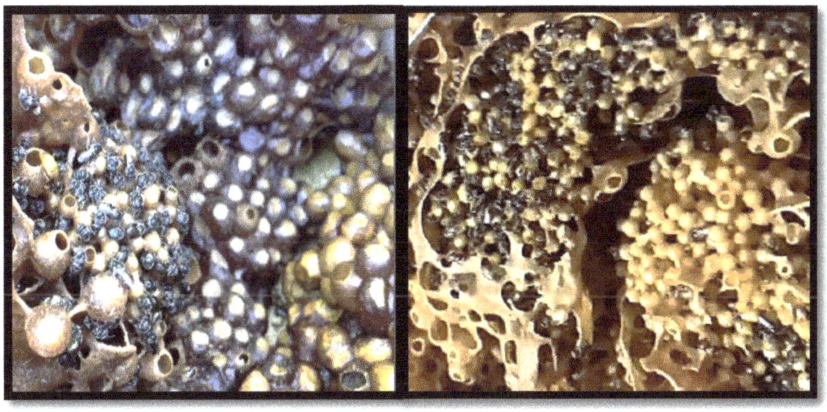

Figure 95 Brood of A. australis, Left: Brisbane; Right: Elenora, SE Queensland

The brood cells image from Peter Davenport, Elenora, SE Queensland, shows the same characteristics.

Images below are from Matthew Middleton, who facilitates small Australian native bee workshops at his Mooloolah Valley bee shed and garden.

Figure 96 Austroplebeia Australis brood from damaged log

The advancing front of *Austroplebeia Australis* hive; callow (baby bee emerging from its cell) in the left photo, *Austroplebeia Australis* brood from a damaged log in the middle *Austroplebeia Australus* honey pots in a boxed hive in the right photo[7].

[7] Editor's note: these (on the far right panel) look too much like food pots, particularly pollen

8.1.2. For *Austroplebeia cassiae,* the involucrum is like an arabesque lace blanket. It may cover the whole brood cell clump or just a portion of it. The h/w ratio = 1.227 ± 1% & the c/w ratio = 0.212. Cells are clustered types.

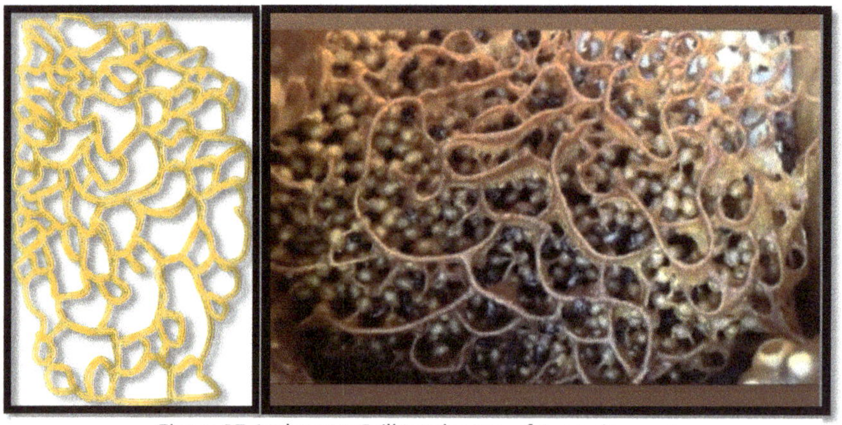

Figure 97 Arabesque Grill Involucrum of A. cassiae nest

8.1.3. In *Austroplebeia cincta,* the brood is generally small, indicating a low average nest population. Cell to cell is flat connectors at cap brim to next cell body, and the c/w ratio is 0.4 – 0.6. Cell to involucrum is tube type where c/w ratio = 0.189. The avg. h/w ratio = 1.143 ± 1%. The brood cell shape is uniformly Prolate Spheroid with no distinct difference in size or colouration between larval and pupal cells.

Figure 98 Cell to cell adhere with flat connectors at cap brim to next cell body. Cell to involucrum is tube type.

Figure 99 Aistropleneia cincta @ PNG Biomass Switpela bi Hani project by Matthew Middleton

Austroplebeia hives have much lighter propolis than *Tetragonala* hives. The beautiful interior in the picture of the *Austroplebeia cincta* hive is from PNG, and it was taken during the PNG Biomass Switpela bi Hani project (Matthew Middleton).

Figure 100 Austroplebeia cincta in the Wasur National Park, Merauke, Papua

La Hisa, a ranger in the Wasur National Park, Merauke, Papua, Indonesia, share this image of an *A. cincta.* The maculations on the scutum are very recognizable, and the nest structure is the same light cream colour.

8.1.4. *Austroplebeia essingtoni*

Referring to Article 25 of the Aussie bee website (https://www.aussiebee.com.au/abol-current.html) includes nest entrance and brood photos. The images by Tim Heard shows the Larva cells approximate ratio of h/w is 1.177, and the Pupal cells approximate ratio of h/w is 1.367. The values here are approximate as the image was low resolution, making it impossible to assess the connectors.

8.1.5.Austroplebeia magna

Referring to Article 25 of the Aussie bee website (https://www.aussiebee.com.au/abol-current.html) includes nest entrance and brood photos. The images by Aung Si have the Larva cells h/w ratio of 1.01 ± 1%, the Pupal cells h/w ratio at 1.311 ± 1.5% and the connectors c/w ratio at 0.168.

8.2 *Lisotrigona* brood structure

8.2.1 Lisotrigona furva brood cells

Interestingly, another configuration was found from Thành Phố Bến Tre from Vietnam. It has a nest

Figure 101 Lisotrigona furva
ref: https://www.youtube.com/watch?v=t6192gxldcs

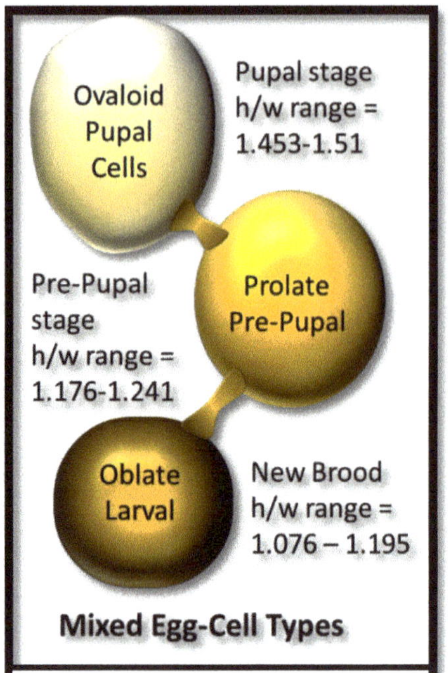

Figure 102 Mixed Oblate & Prolate Spheroids + Ovaloid shapes cells

similar in structure to the mixed brood cells combination, and it is of the stingless bee *Lisotrigona furva*.

Lisotrigona furva.

Connectors: The connectors here are fairly consistent throughout. The connectors (c) to avg. cell width (w) ratio ranges 0.142 – 0.16 in all stages, i.e., the larva, pre-pupal, and pupal stages.

8.2.2 Lisotrigona chandrai brood cells

Figure 103 Lisotrigona chandrai Brood cells

Chapter 9

9.0 Discussion

9.1. *Geniotrigona thoracica*

A decade ago, I was watching my first *G. thoracica* colony expansion. They were expanding into the 2nd tier of the topping box. As usual, they made lots of struts and pillars first. With more struts, I was expecting lots of honey pots. However, I saw brood cells, and it was the start

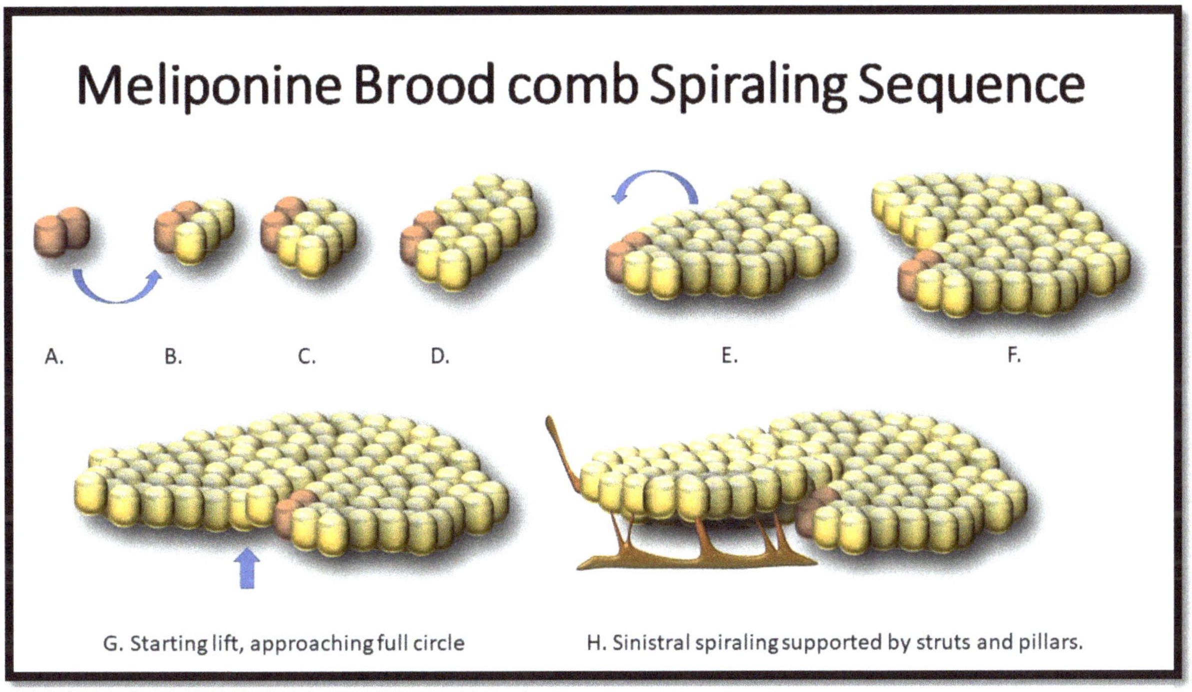

Figure 104 Meliponine Brood comb Spiralling Sequence

of a brood chamber build.

In Figure 104 A. The initial couple of cells. B. they attach more to the right of each. C. & D. the build-up was similar, always adding to the right. E. when the 'wafer' was sufficiently wide (wide to them, I guess). F. They start orbiting the initial cells (in a darker orange shade), never adding any cell to the left of the initial ones. G. as they approached the full circle, there was a perpendicular upward lift (as though avoiding to contact the left of the initial cell. H. the lift came about and was supported by the struts and pillars attached to the involucrum and batumen. What resulted was a sinistral spiral in all *G. thoracica* spiral broods that I have seen. Never dextral.

I recollect a video; a group of us watched how a *Lepidotrigona terminata* collects Jackfruit gum and loads her hind tibia. Someone in the group said, "Wow, these bees are so Arabic". I looked puzzled. "Huh!?" The explanation = Look at how each time the bee loads her right tibia first, then the left. Then take off with the left hind tarsus last to leave the ground.

Is that what some people take from such an intriguing bee behaviour video? As it turns out, we could not rebuke as we watched *Geniotrigona thoracica* that did many activities right-handed first. See, in the Arabic tradition (as we were taught young), we only touch our food with our right hands (even left-handed or "southpaws" may do everything else with the left but must eat or touch food with the right... (handicaps excepted). We leave our domicile, dwelling or place of work or any journey always with the right foot first! We serve or hand over anything or even point with the right hand.

We find both dextral and sinistral whorls occurring almost equally among those spiralling. I find it intriguing that the whorls do not matter, whether in the Northern or Southern hemisphere. In entrance tube construction and brood spiral, *Geniotrigona thoracica* is always sinistral.

Figure 105 Anti-clockwise entrance tube.

9.1.1 Right-handedness Trivia

The Japanese Bushido code says that one should always wield the weapon with the right hand. That's why they don't shake hands as we do, for fear of a concealed weapon in the right sleeve. They bow to each other. So, the other day I was chatting with a Biroi worker about this right-hand and left-hand thingy. She said Filipinos were Spanish orientated earlier and Americanised later. They start eating with a knife in the right and fork in the left and change the fork to the left after cutting the food. So, the right-hand fork goes to the mouth. Then what about the first Biroi in PNG? They were German earlier, so Friese (an Austrian German) named her after the collector Lajos Biro (a Hungarian). Then what about Sulawesi... oh, they're Dutch. How about *carbonaria* in Australia... oh, they were British, a bit confused; Europeans bring the left fork to the mouth.

Hair whorl is occasionally used to illustrate basic genetics, and it gets more attention because some studies have suggested that counterclockwise whorls are more common in left-handed people than in right-handed, and other studies have suggested that counterclockwise whorls are more common in gay men than in heterosexuals. https://en.wikipedia.org/wiki/Hair_whorl have significance in Animal behavioural theories.

Snails have coiled shells, and most coil in the right-hand direction: if you look at the shell with its tip pointing up, the opening is on the right side. The sinistral whorls do not occur in all species and are rare. The Hindu reveres them, and the Hindu priests usually keep one in their temple.
The shell swirl reflects the handedness of the snail's internal organs. In some snail populations, the reversal resulted in the evolution of new species. In other cases, left-handedness - a shell whirl coiled counterclockwise - might mean the snail never even reproduces.

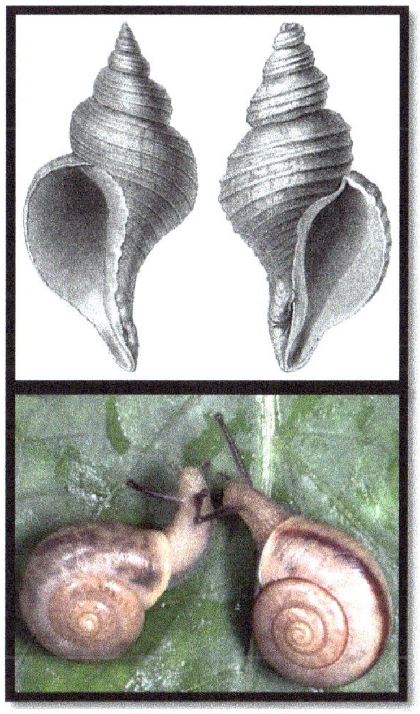

Figure 106 Sinistral and dextral coils of molluscs.

9.2 Honey Woes

Almost all Stingless Bee Honey sold in India is "*iridipennis* honey". Even in Nagaland, the full comb layered brood is sold as "iridipennis honey". (See video: https://www.youtube.com/watch?v=OhQ4hj6wUoE&t=34s. I have communicated with several beekeepers over the past five years, and they mostly say they have *iridipennis* honey. They know a few species exist in India, but they presume it's all the same and just mix up and bottle them. After all, honey is honey, and people just want honey.
In The Philippines, Stingless Bee Honey is either "*biroi* honey" or monggo-monggo regardless of species. Only of late, the beekeepers get an inkling of *sapiens, laeviceps* or *iridipennis*. Still, they can't differentiate the species, so lump everything together, and it is easier to meet the market demand for qty; in the back of my mind, if you are going to mix, why bother segregating *biroi*? What I mention here is not indicative of all the beekeepers, but many I met mix them up.

In Indonesian Wallacea, like in Lombok, both sapiens and clypearis honey is mixed and sold to other islands, much sought after for their sweetness. In Sulawesi, *Tetragonula* honey is secondary because the incisa honey supersedes the *Tetragonula* production. However, they have their *T.* nr. *biroi* honey all other *Tetragonula* species are not seriously harvested and commercialized. Sumatra similarly has bigger producers like *itama, thoracica* and *apicalis,* so *Tetragonula* species honey is taking a back seat. In Java, especially Central Java, everything is *laeviceps* to them, and the beekeepers are getting meagre amounts from the native Javan *Tetragonula*. So, I have met a few that I know of keeping more than a couple of thousand hives to get as much as possible. They cannot be bothered by their species. After all, one said, they are *laeviceps* (*induk acak* - Indonesian for clustered type brood). They have started bringing *T.* nr. *biroi* from Sulawesi.

Borneo is more diversified with *Heterotrigona, Homotrigona, Tetrigona* and *Lophotrigona,* all the "big guns", so *Tetragonula* is not cultured. With Brunei's influence, *fuscobalteata* culturing is fetching some good returns. The price is high for the Indonesian when they compare with Brunei demands.

In Thailand, I cannot interact much because I don't speak the language, but I gather from Dr. Anchalee Sawathum and Dr. Bajaree Chuttong that they culture a few *Tetragonula* and *Lepidotrigona* species besides *Heterotrigona* and *Tetrigona.* In 2017 we had a Honey ranking competition in the Malaysian Genome Institute. The Judges were International combined with NMR analysis. Anyway, the issue is. Two samples of *Lepidotrigona* won 1st and 2nd, and *Tetragonula testaceitarsis* won 3rd place (Dr. Anchalee brought all three from Thailand. From then on, *Lepidotrigona terminata* and *T. testaceitarsis* is a favourite in Malaysia.

In Malaysia, the beekeepers learn to recognize the bees by their nest entrance. Granted, it is not the best of criteria, but at least they harvest and accumulate their honey segregated by nest entrance, as they learn other methods of identifying the species, only then do they name or brand their harvested honey.

Right now, I find that Vietnam is vastly picking up on Meliponiculture and the variety of species and complex of species is difficult to recognize. The nest entrances are nothing much to go by, judging from the videos I have seen. The hives are new, so the entrance tubes have not developed. However, it is good that they can differentiate *Tetragonula* from *Lisotrigona.*

9.3 *Homotrigona (Lophotrigona) canifrons* - rarely seen hires images because of their very aggressive nature, and beekeepers find it difficult to examine its brood[8]. Ovaloid brood cells h/w =1.32 -1.35. Images by Muhammad Ashraf of Johor, Malaysia.

[8] Editor's note: Nesting also occurs in rock cavities, as I saw just outside Kuala Lumpur in limestone hills.

They have a peculiar way of connecting the cells like clusters of grapes and the thick connectors like grape bunch pedicels.

Figure 108 Homotrigona (Lophotrigona) canifrons

9.4 *Wallacetrigona incisa*

The *Wallacetrigona incisa* brood is rarely photographed because they are only found in certain remote regions of Sulawesi. Those who culture these bees are more motivated for honey production than studying their nest architecture. Irregular segments layered like terraces but appeared cascading in some sections.

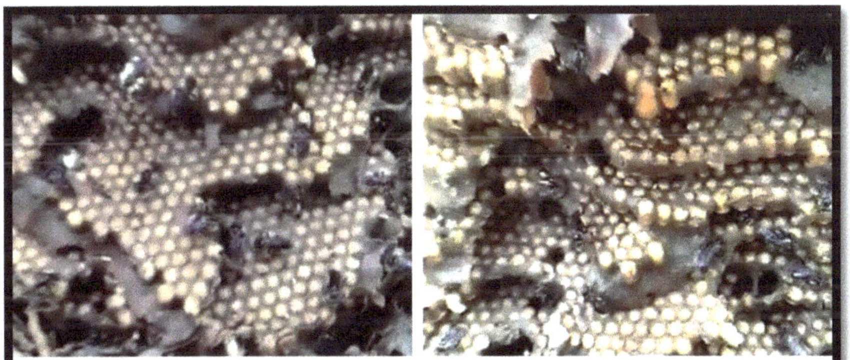

Wallacetrigona incisa Images by Ismayunto, Baruga, Kendari, SE Sulawesi

Figure 109 Irregular segments layered like terraces but appeared cascading in some sections of the Wallacetrigona incisa brood.

This situation may be due to the vertical type box hives used by Sulawesi beekeepers. Elongated Cylindroid brood cells with h/w = 1.6.

W. incisa Queen antennation - Images by Hassan Asri, Makassar. Sulawesi

Figure 110 The Wallacetrigona incisa queen here has resemblance to Geniotrigona thoracica.

9.5 *Heterotrigona (Platytrigona)* cf. *flaviventris* (Friese, 1908 [1909])

These images are from La Hisa, Wasur National Park, Merauke. It was found in a rotting log. Ovaloid brood cells h/w =1.263 -1.4.

Figure 111 Heterotrigona (Platytrigona) cf. flaviventris (Friese, 1908 [1909]) found in a rotting log

9.6 Inference

As all bees have brood cells, I deduce that the cell's shape classification format can be applied similarly to semi-combs or other cluster formations and other Meliponine genera. For the semi-combs, a side view is necessary to ascertain the cell shape and, more importantly, the average cell height to cell width ratio,

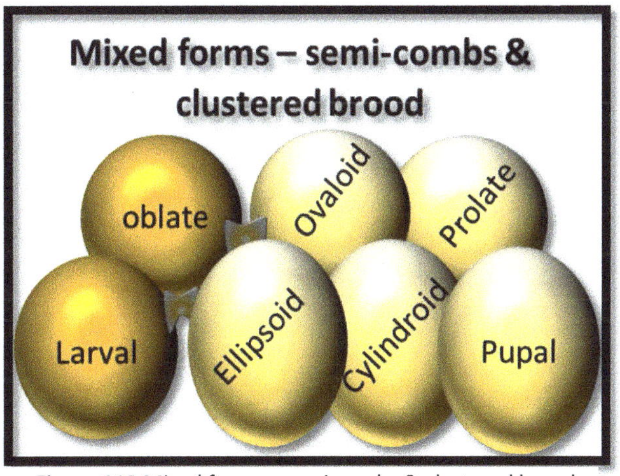

Figure 112 Mixed forms – semi-combs & clustered brood

I am confident we will acquire a new supplementary criterion for determining the species identity with sufficient data.

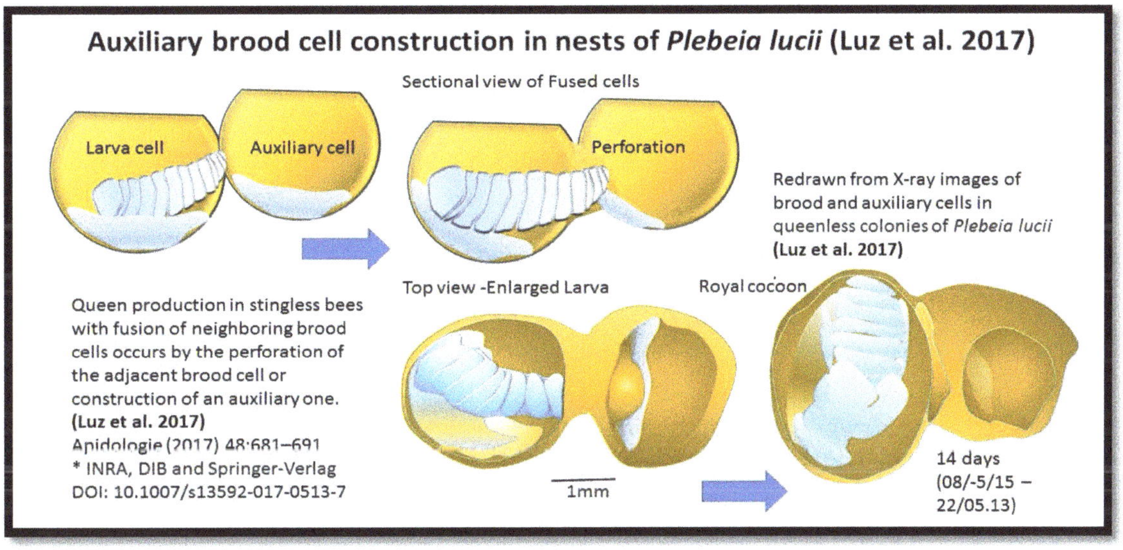

Figure 113 Auxiliary brood cell construction in nests of Plebeia lucii

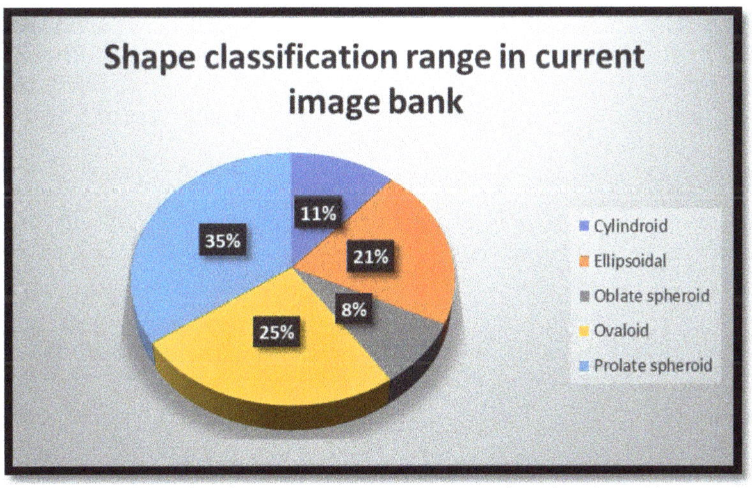

Figure 114 Shape distribution among 242 Tetragonula colony brood images.

End of Part 1

Part 2

Geometry and Colours of Meliponine Brood Cells

Chapter 10

10.0 Prelude to Part 2

I am reminded that I attended a Technical School almost five decades ago doing my O-Levels and A-Levels (British education system then in Malaysia). I took a course in Engineering Science. One of the early subjects was Automobile Engineering. In that subject was a test where I had to disassemble a 4-stroke engine and then put it all back together within a fixed time frame. I did it within the stipulated time and was quite proud of it when others could not do it. Surprisingly, I did not get a passing mark because the invigilator found a couple of bolts left behind (among the tools) at the corner of my work table at the end of it. I failed that test, and I decided to major in technical drawing. (Although, now I realize Scientific illustration of insect's anatomy is a different ball game)

That's how I felt when faced with the numerous images of *Tetragonula* brood. I thought I had it all covered. Relating ALL (old and new) the *Tetragonula* species to the many different brood cell shapes and forms. I may not have passed this test because a couple of species were not matched yet to a couple of brood forms I recently found.

Classification is not a science in itself but a means of description and communication. It is a step towards understanding things in the natural world. We often overlook classification criteria to take the easy way out. The issue at hand is the classification of honey associated with species or some consistent characteristic. Tremendous interest in local stingless bees is already seen from the increase in beekeepers and producers of the highly appraised and valued "kelulut" honey produced by meliponines, likely including *W. incisa* (Rasmussen et al. 2017). Thus, there is a concern that the kinds of honey harvested may be having mixed sources of different species. The beekeepers must be educated to recognize the species one way or another. On the other hand, too detailed or too much categorization leads to confusion[9].

[9] Editor's note: you might add that the initial impetus to perform whole-genome DNA sequencing was to be able to identify 'counterfeit fish' in the market. Their flesh in fillets was too easy to confuse.

These brood cell shape and brood form classifications are preliminary and are not diagnostic features or characteristics. Until they are established, stingless beekeepers may segregate their colonies by these categories and not have all species of honey mixed during harvesting.

Too often, new names are given to something that has already been named. However, this book deals with the associated brood images' names. It is not the job of this author to rationalize the synonymy or other combination of names for a newly named species even though they may exist as an earlier name in another region. This literature does not replace the scientist's job but merely allows readers to be more scientifically literate.

A decade ago, with the paper, The Remarkable Nest Entrance... Nesting in Limestone Cavities, Bänziger, et al.; JKES, 2011, many Malaysian beekeepers (me included) started harvesting stingless bee honey and grouping them by the nest entrances. We could not even pronounce the names correctly, and what more to memorize the numerous varieties. We had local names after the entrance shapes, like the 'Buffalo's back-end' for *thoracica*[10]; 'sun-rays' or 'sunburst' for *testaceitarsis*; 'trumpet' for *itama* and so on. We, beekeepers, do not like to read much, so the images from that paper became the benchmark for recognizing the bees then.

I thank Dr. David Roubik for reminding me how tropical heat and gravity affect the nest structure internally and externally with that intro. I recollect my first article on Meliponary management, how the dipterocarp dependent bees have their entrances distort and droop due to heat and gravity when provided alternative resin sources. The brood structure internally would be affected in some manner.

"Basic differences will always occur in the tropics, and sometimes heat, and the cerumen in pillars. Things do respond to gravity. Pillars deform, broad cells coalesce, and brood connected directly to a wall is the only kind (among the 'cluster type' cells) that can be uniform and tight in the arrangement." (Roubik) He further mentioned, "And anywhere that a larva, pupa or imago has died and needed to be removed, there will be a gap that will be filled, perhaps with a pillar or connective, or maybe not. Holes in brood comb and strange irregular gaps in cell clusters are likely due to the same cause- brood death and hygiene. While I appreciated these suggestions, I received a couple of images from Dr. Pham Hong Thai (contributed by Tran

[10] Editor's note: That's remarkable. In Costa Rica, the common name for one of the most common stingless bees, which I have studied well (*Trigona fulviventris*) is that of 'cow's back end'-and they are not referring to adult bee morphology (but the nest entrance).

Van, the same contributor for the ball-n-spoke brood, who, by the way, is in South Vietnam, not N Vietnam)

Thanks to Dr. Shanas. S., Kerala Agricultural University, Kerala, India, I now have the classification of brood cells of the three species he named a couple of years back.

10.1. *Tetragonula (Flavotetragonula) calophyllae* Shanas and Faseeh, n. sp.

Figure 115 Tetragonula (Flavotetragonula) calophyllae Shanas and Faseeh, n. sp.

Prolate Spheroid Larval h/w = 1.132 ± 1%; Ovaloid Pupal cells h/w = 1.387 ± 1%

10.2. *Tetragonula (Tetragonula) perlucipinnae* Faseeh and Shanas, n. sp

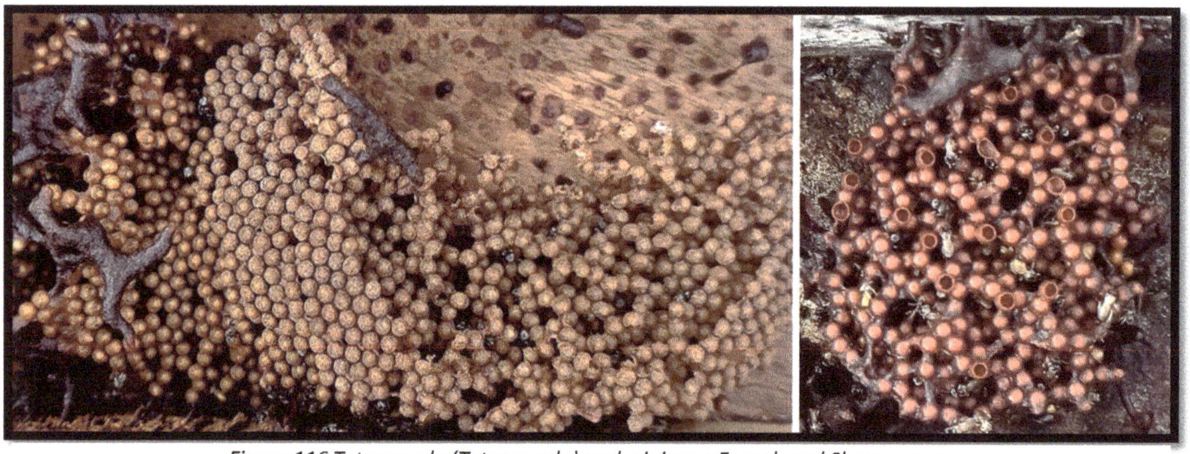

Figure 116 Tetragonula (Tetragonula) perlucipinnae Faseeh and Shanas, n. sp

Prolate spheroids Larval perforated clustered brood form h/w = 1.12 -1.151;

Ovaloid pre-pupal perforated semi=comb brood form h/w = 1.229 ± 2%

Pupal full comb layer h/w = 1.18-1.294 brood - Can be taken apart in layers carefully (Shanas S.).

I have seen this type of form in brood images from Vietnam.

10.3. *Tetragonula (Tetragonula) travancorica* Shanas and Faseeh, n. sp.

Figure 117 Tetragonula (Tetragonula) travancorica Shanas and Faseeh, n. sp.

The Ovaloid brood cells colour varies a lot, from golden brown to cream. More importantly, it has features like a graduated brood (similar to the *T. mellipes* graduated brood), sparse larval brood h/w range 1.164-1.187 in the middle and dense clusters of cocoons h/w range = 1.227-1.297 at the periphery. This graduated brood feature is more prominent in the cream-coloured brood.

10.4 *Tetragonula* cf. *zucchii*

I found two more types of brood form by Tiny Thomas of Kerala, India, on his fb website "Tiny

Figure 118 Tiny Thomas of Kerala, India - T. cf. zucchii

Bees' '. I compare this first one with *Tetragonula* cf. *zucchii* I observed in Langkawi Island, Malaysia.

Ovaloid h/w = 1.352 ± 2% https://youtu.be/A8HvfKvRwrk. Video of a rare species of Stingless bees found in the northern part of our state. Honey pots and 'egg cups' are made of white propolis. Honey produced by them does not have any colour. Bees are very sensitive. They come out of their hive by around 10.00 am and forage only when there is a clear sky. (Tiny Thomas)

10.5 Uncommon species - *Tetragonula* sp.

I have not determined the species in this second type or speculated close to it.

Figure 119 Tetragonula sp. from Tiny Thomas of Kerala, India

Prolate spheroids h/w = 1.1243 ± 3%; Ellipsoid Q cell h/w = 1.437

10.6 Recapitulation on Brood forms

The rundown for *Tetragonula* brood forms I gathered thus far is this:

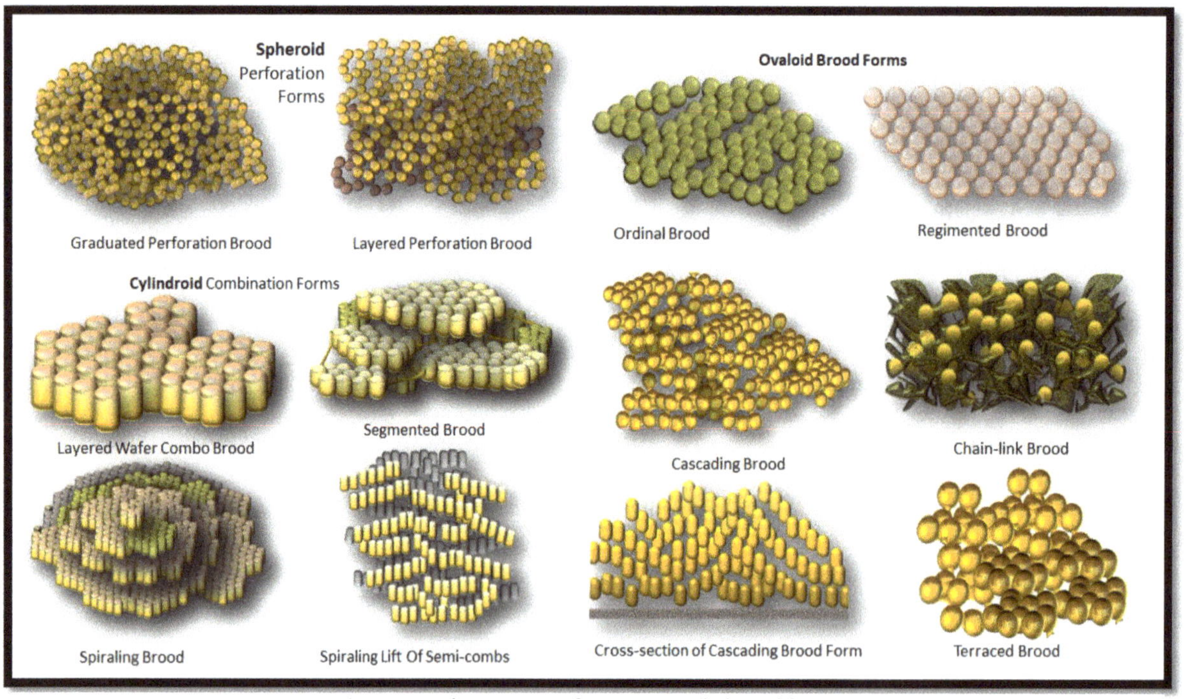

Figure 120 The summary for Tetragonula brood forms

So, I checked my work table for any 'bolts' left behind, and true enough, along came to a spider and dropped a couple of brood images on my lap(top). It is a feral hive in a tree trunk hollow from Vietnam.

Figure 121 Combo of Ball-n-spoke brood & semi coms segments.

What?? When is this going to end? I thought to myself. I started looking intently at the bees - the resolution was not the best, but I could make out that it did not have that projecting scutellum, propodeum was visible, scutum was shiny and no hair bands. Goodness, it's not a *Tetragonula* species! (Recollecting what I learned from Prof Deborah Smith in the AAA conference in 2012)

10.7 *Lisotrigona* Brood forms

I started reviewing my notes and papers on Meliponini from Vietnam. The Engel 2000 key on *Lisotrigona* fascinated me the most. So, it does not look like an *L. furva* brood. I have seen *L. cacciae* brood image from E. Kalimantan, which does not resemble this. I started looking for the signature yellow clypeus of *L. carpenteri*. If given a chance, the poor image resolution is of no help, but I would still compare this to *L. carpenteri*.

Having technical drawing skills, I drew the top layer of cells (I presume they are larval cells), which I got.

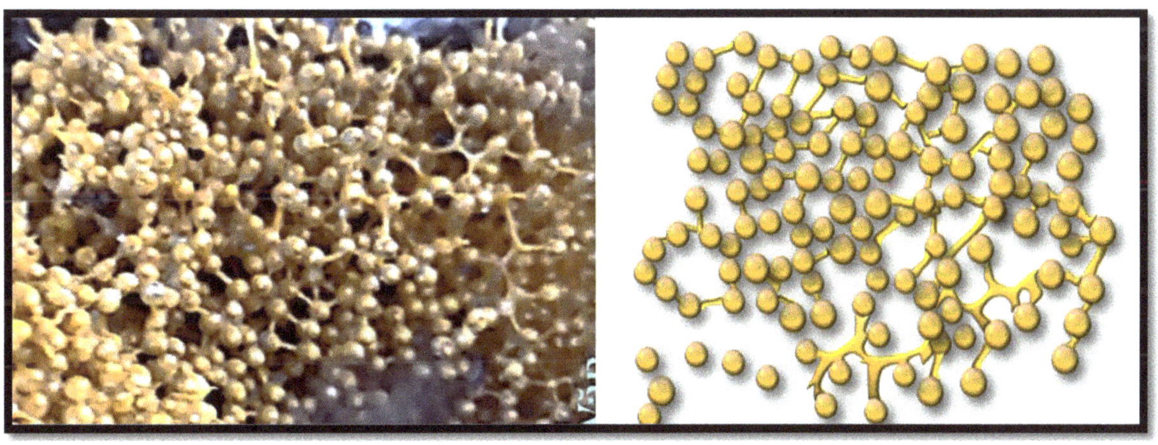

Figure 122 Ball-n-spoke brood of L. cf carpenteri or sp.

Instantly I started reminiscing about my school days of chemistry lessons on molecule models. It sure looks like the ball-and-stick models in chemistry.

I decided to call it the Ball-n-spoke brood of *Lisotrigona* cf *carpenteri* or sp.

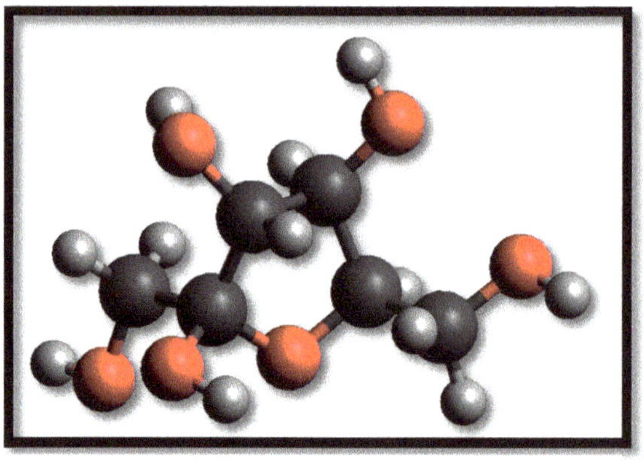

Figure 123 ball-and-stick molecule models.

10.8 Regimented Brood

I thought I was done covering all the different types of *Tetragonula* brood cells. Well, what do you know, there's one more?

Figure 124 The cocoons are equally spaced in uniform rows and columns; I call it the Regimented Brood - Tetragonula sp.

The cocoons are meticulously arranged, seemingly equally spaced in uniform rows and columns; I call it the Regimented Brood. The larval cells are randomly clustered with Ovaloid to Ellipsoidal shapes, but the cocoons are Cylindroid and so well arranged you could not have set them better yourself manually[11]. I can't determine the species yet.

[11]Editor's Note: the equal spacing comes from 4-piece stabilizers. The mass will not shrink when the cerumen of cells is removed. And it is the placement up against the wall itself- which has nothing to do with the pupal stage per se. the pupae had egg and larval cells in the same place and in the same formation beforehand. It is attached to a wall and each cell to four others. Those outside this layer MUST be configured differently.

10.9 Curved brood form combinations

These are the images; many thanks to Dr. Pham.

Figure 125 images contributed by Tran Van via Dr. Pham Hong Thai

The first one looks like a comb brood with the strange irregular gaps that Dr. Roubik mentioned. The second seems like a marriage of a perforated brood with the ball-n-spoke brood.

I decided to look closer at the comb-like brood. It is not a flat wafer comb but became coalesced to a 'compact and tight' brood relative to the cylindrical shape of the tree branch cavity. It appears similar to the *T. iridipennis*? brood from India (see below)

Figure 126 T. iridipennis? Brood from India.

On closer inspection, the irregular gaps in this one from Vietnam are the segregation of different batches of the brood. It became obvious when I drew lines following the rows of cells.

Figure 127 different batches of brood, obvious from lines following the rows of cells.

Figure 128 wafer comb species (T. nr. biroi) with irregular gaps.

I compared this with a known wafer comb species (*T. nr. biroi)* to look for irregular gaps.

In the first image, the segregation of batches is more pronounced with clearly defined quadrants, and each batch of quadrants has its irregular gaps. I agree with Dr. Roubik that these gaps are where a larva, pupa or imago has died and needed to be removed. This image is a top view of one cultured in a box.

Figure 129 clustered type T. laeviceps in a cylindrical hollow of a bamboo node or tree branch

10.10 Cylindrical host cavity

I then looked at the cylindrical host cavity. The next image is a truly layered brood comb in a cylindrical tree trunk hollow. Full combs and semi-combs in a cylindrical container are typically as below. You may not see the 'strange holes' when dissecting the cylinder. Individual cells are arranged vertically (y-axis)

And for clustered types like *T. laeviceps* in a cylindrical hollow of a bamboo node or tree branch are like these (See Figure 129).

It is like the cavity shape moulds them. However, the individual cells are arranged horizontally (x-axis). We only see the cell caps (opercula) as a comb brood. In some, the cells are arranged obliquely (z-axis) if in a box as this one below. The young brood does exhibit some ball-n-spoke characteristics.

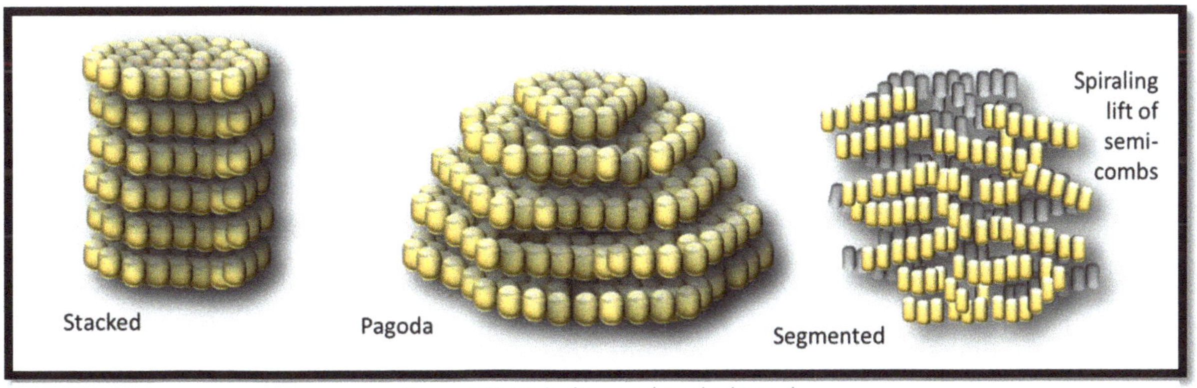

Figure 130 Typical Layered stacked Brood

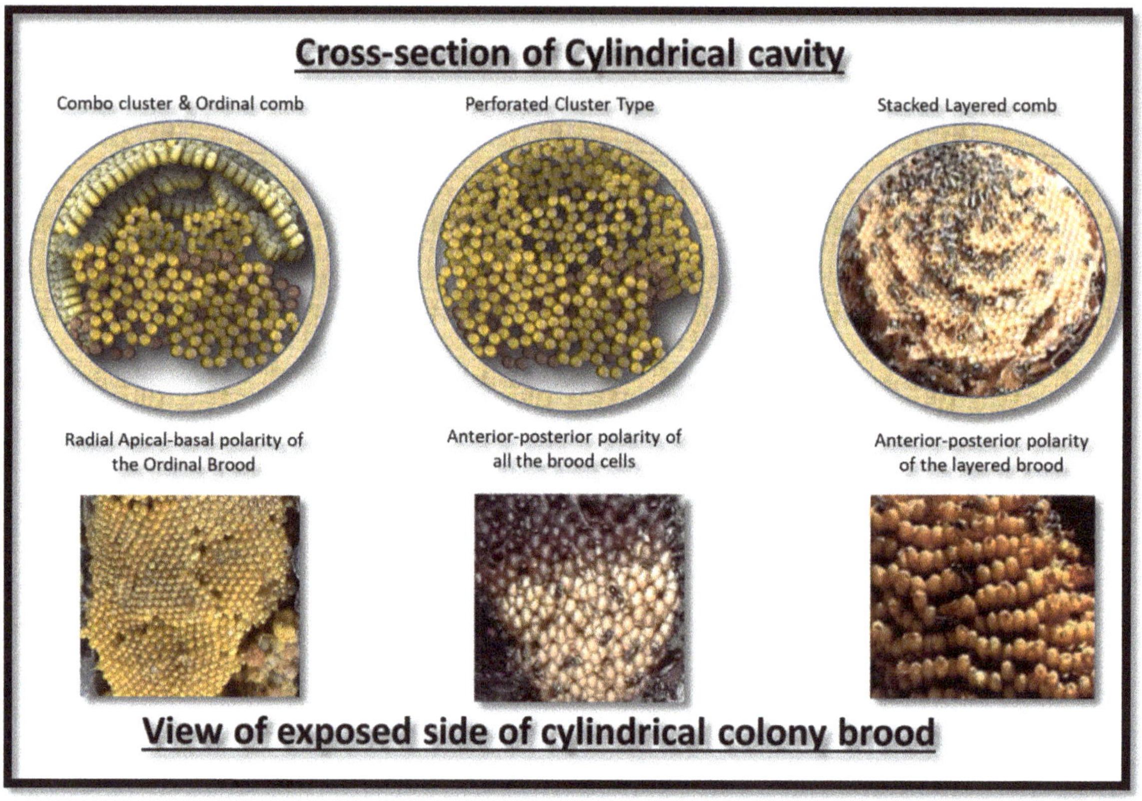

Figure 131 Sectional and side views of various brood forms in a cylindrical cavity.

On the top left is the cross-section of semi-combs with clustered type broods of species like *Tetragonula drescheri*. The middle shows clustered type of *T. laeviceps*, and the right image is of the layered comb brood of *T. carbonaria.* or *T.* nr. *biroi*. I aim to explore other types of broods and how they appear in different cavities, especially in branch crotch cavities with a Y shape hollow. How would the spiral form adjust in such a cavity?

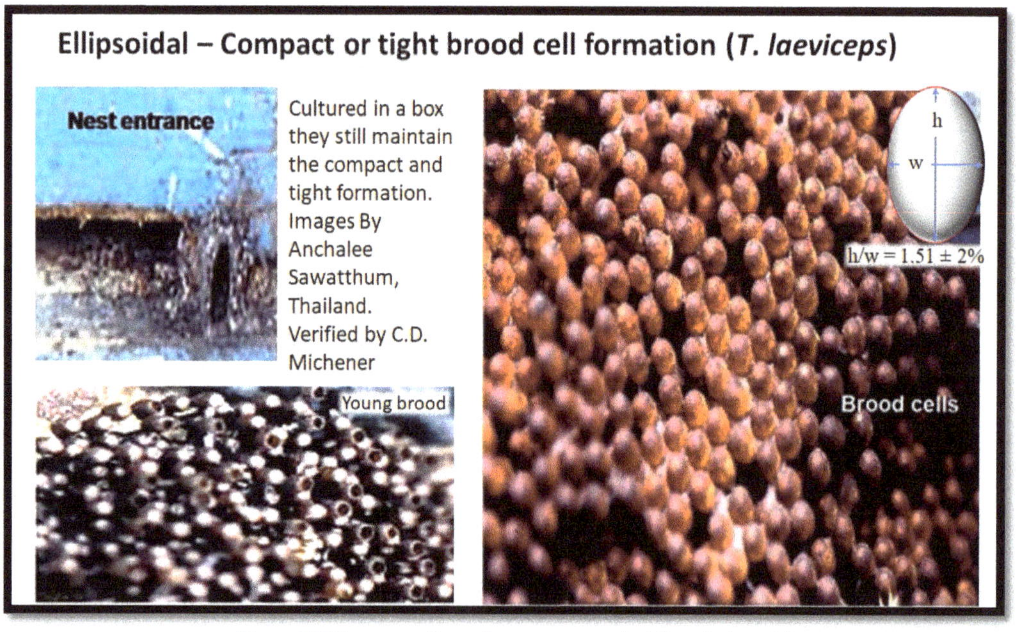

Figure 132 In some, the cells are arranged obliquely (z-axis)

10.11 Chirality significance in Brood form

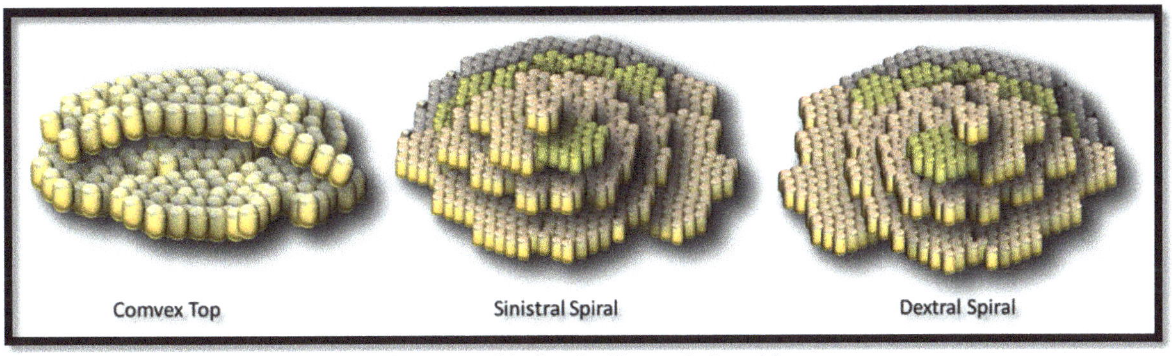

Figure 133 Chirality significance in Brood form

Chirality[12] may not be significant because different form combos and handed spirals occur randomly in the same nest structure. The forms or combinations of forms appear without any indicative building block. There are up to four different forms in the separate quadrants without any symmetry of form (See Figure 83). The quadrants in the brood cell structures are independently random (although other influencing factors are not fully studied and understood).

10.12 Miscellaneous Brood images from Vietnam

It appears that in examining cluster type and semi-comb type of *Tetragonula*, we may infringe into brood forms of *Lisotrigona*. This issue is evident in many images from Vietnam.
One shows a curly segmented brood where the singular bee looks like a *Lisotrigona* sp. Another has a similar (curly segmented brood) in combination with a concave layered comb brood. The next one is a cascading brood combined with an inclined Ordinal brood.

Figure 134 a cascading brood in combination with an inclined Ordinal brood.

[12] 'AHJ: Chirality' is derived from the Greek word for hand. A chiral molecule is distinguishable from its mirror image just like the left hand is distinguishable from the right hand.

Figure 136 Here is a similar brood in combination with a concave layered comb brood.

Figure 135 a curly segmented brood where the singular bee looks like a Lisotrigona sp.

10.13 Initial Recourse

However, I still have other questions that I truly hope someone can help me with:

1. How did the Meliponini brood form evolve? Was it from clusters to combs or vice versa? Perhaps those studying the evolution of the Meliponini brood have some insights.
2. Is there any fossil evidence or traces of fossilized Meliponine broods?
3. Has the chirality of the sinistral and dextral spiral broods had any significance in their development?
4. Why do some brood cells have anterior-posterior polarity, some apical-basal polarity, while some are obliquely arranged? What determines whether its arrangement or brood was coalescing is purely influenced by cavity restrictions, species diversity, or other factors. Some species create external extensions (with their involucrum) in situations of restricted cavity size.

There is a saying, "Never memorize anything you can look up!". I have been looking up answers to these questions and have not found the answers yet. Nevertheless, this book may serve as a reference for those who do not wish to memorize whatever information I have compiled herein.

Chapter 11

11.0 Of Queen cells and Queens

11.1 *T. iridipennis* & *T. ruficornis*

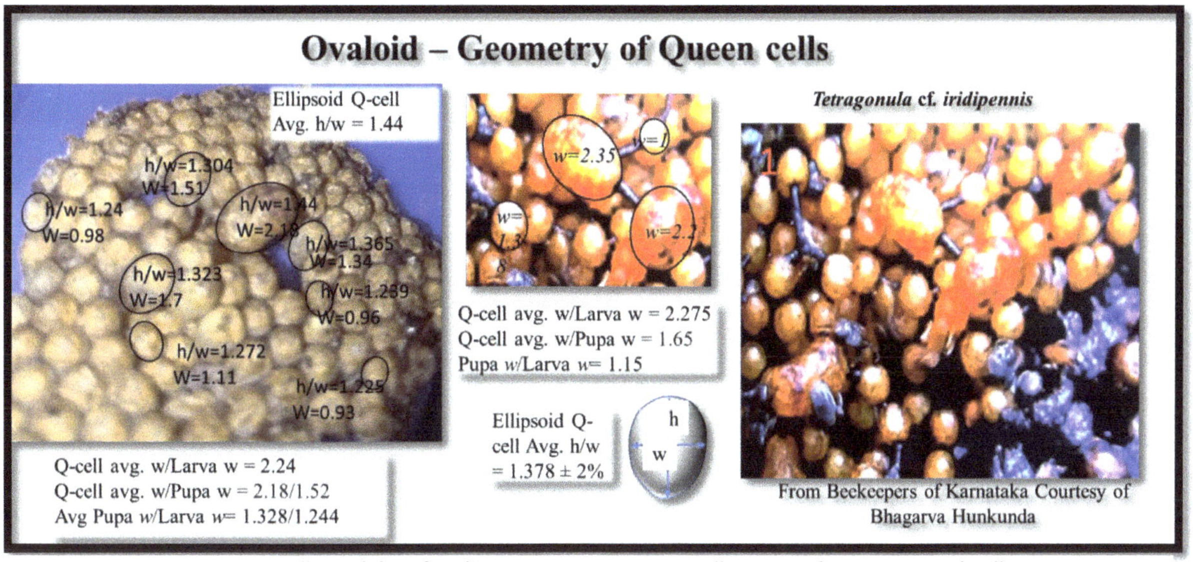

Figure 138 Ellipsoidal T. cf. iridipennis average Queen cell = 2.575 of average Larval Cell

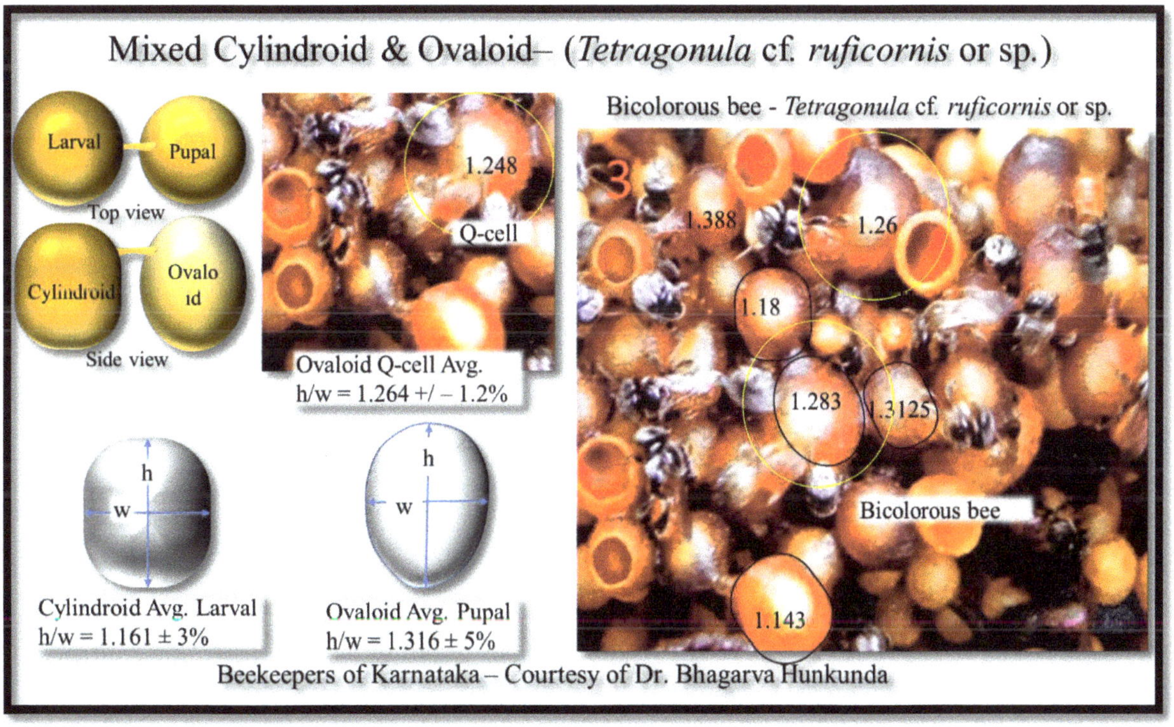

Figure 137 Ovaloid T. ruficornis average Queen cell = 1.506 of average Larval Cell

1st sample *T.* cf. *iridipennis* - Ellipsoid Q-cell Avg. h/w = 1.378; Q cell avg. w/L w = 2.275; Q cell avg. w/Pupa w = 1.65; P *w*/L *w*= 1.15; - **2nd sample *T. iridipennis*** Q-cell Avg. h/w = 1.44; Q-cell avg. w/Larva w = 2.24; Q-cell avg. w/Pupa w = 1.434; Avg Pupa *w*/Larva *w*= 1.56

T.* cf. *ruficornis Ovaloid Q-cell Avg. h/w = 1.264 +/ – 1.2%; Q cell avg. w/L w = 1.506; Q cell avg. w/P w = 1.365; Pupa *w*/Larva *w*= 1.103

11.2 *T. aff. sarawakensis & T. reepeni*

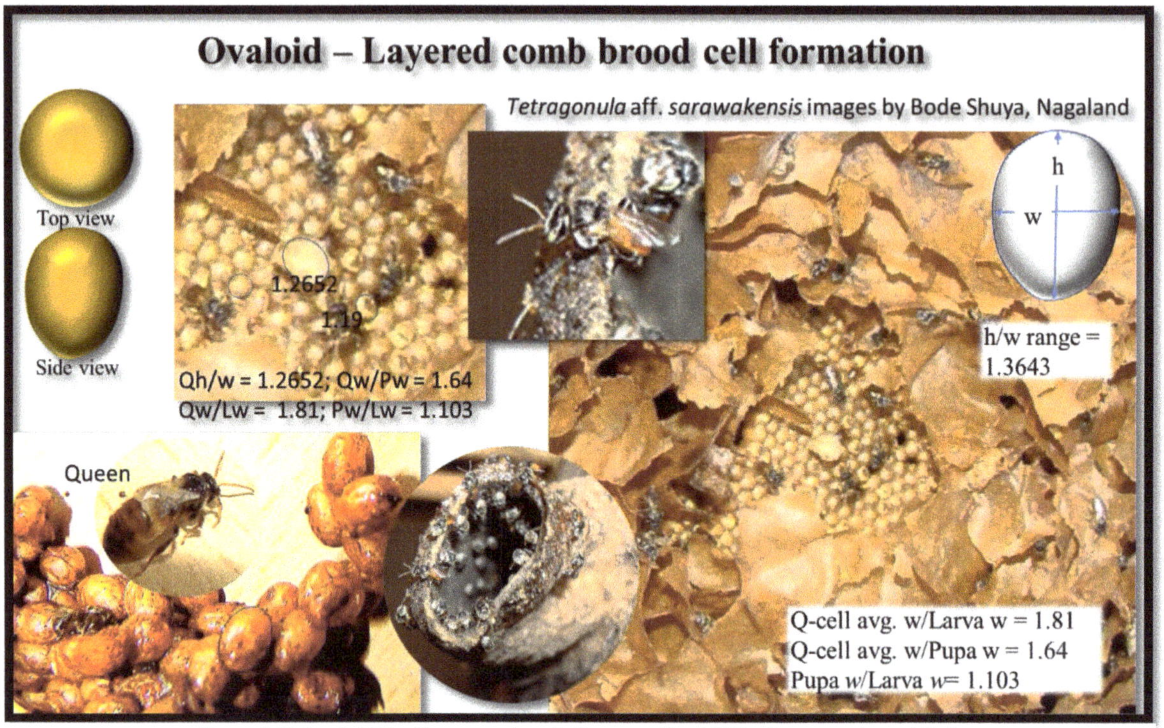

Figure 140 Ovaloid T. aff. sarawakensis Queen cell = 1.982 of average Larval Cell

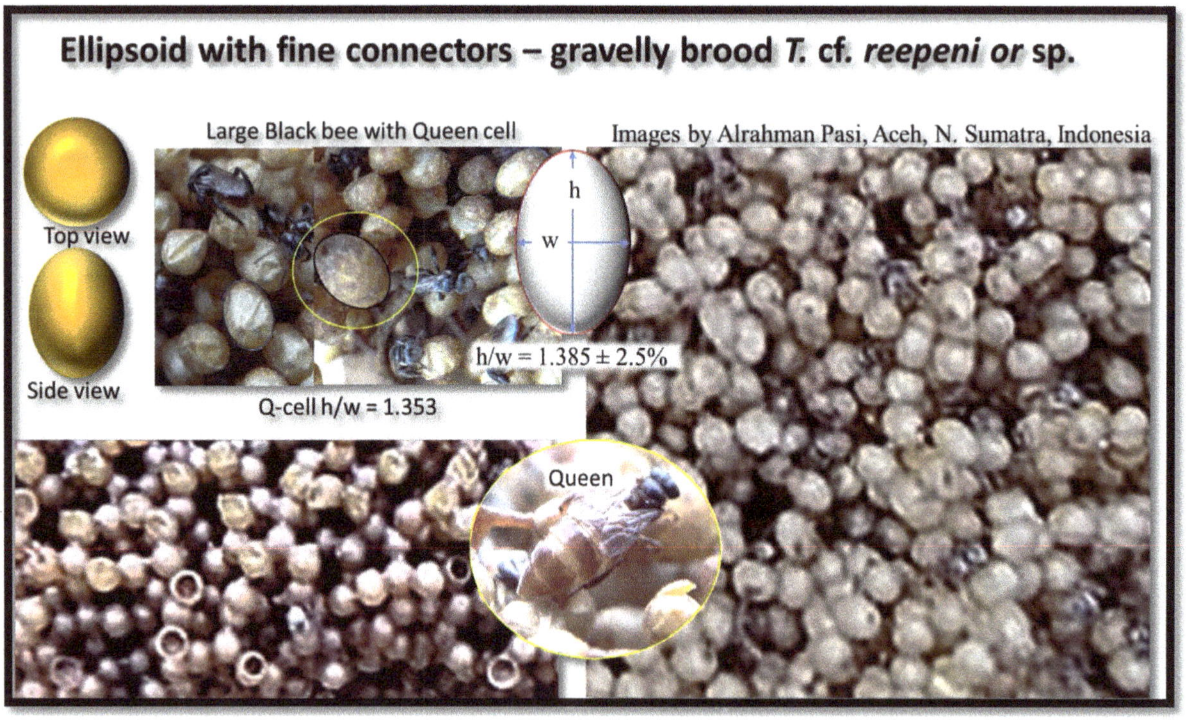

Figure 139 Ellipsoid Tetragonula sp. average Queen cell = 1.982 of average Larval Cell

Tetragonula* aff. *sarawakensis Q cell h/w = 1.2652 Queen cell avg. w/Larva w = 1.81; Queen cell avg. w/Pupa w = 1.64; Pupa *w*/Larva *w*= 1.103;

***Tetragonula* sp.** Q cell h/w = 1.437; Queen cell avg. w/Larva w = 1.982; Queen cell avg. w/Pupa w = 1.8; Pupa *w*/Larva *w*= 1.103

11.3 *Tetragonula* sp. & *T. fuscobalteata*

Figure 141 Ellipsoid T. cf. reepeni or sp. Q cells h/w = 1.353; Qw/Pw = 1.49; Qw/Lw = 1.624; Pw/Lw = 1.09

Figure 142 Prolate Spheroid T. fuscobalteata Q cells h/w = 1.308; Qw/Pw = 1.04; Qw/Lw = 1.241; Pw/Lw = 1.194

Tetragonula* cf. *reepeni Q-cell h/w = 1.353; Queen cell avg. w/Larva w = 1.624; Queen cell avg. w/Pupa w = 1.49; Pupa *w*/Larva *w*= 1.09; ***Tetragonula fuscobalteata*** Q-cell[13] h/w = 1.308; Queen cell avg. w/Larva w = 1.241; Queen cell avg. w/Pupa w = 1.04; Pupa *w*/Larva *w*= 1.194;

[13] AHJ: In *T. fuscobalteata*, the only difference between the Pupal and the Queen cell is coloration where Queen cells are slightly larger (1.04) and beige while pupal cells are cream in colour.

11.4 *Tetragonula* sp., *T. pagdeniformis*, *T. carbonaria*, & *T. laeviceps*

Figure 144 Cuboid T. pagdeniformis or sp. Q cells h/w range = 0.98-1.09; Qw/Pw = 1.614; Qw/Lw = 2.152; Pw/Lw = 1.33

Figure 143 Ellipsoidal T. carbonaria, T. cf. laeviceps (Java) & Tetragonula sp. (Bicol)

11.5 Tabulations

Table 2 Tabulation of Queen cell shapes and Q-cell width to Larval and Pupal cell width ratios.

Species	Cell Shape	Q-cell h/w	Qw/Lw	Qw/Pw	Pw/Lw[14]
T. cf. *iridipennis*	Ellipsoid	1.378	2.275	1.65	1.36
T. aff. *iridipennis*	Ellipsoid	1.44	2.24	1.434	1.56
Tetragonula sp.	Ellipsoid	1.6	1.617	1.32	1.103
T. sp. (Bicol)	Ellipsoid	1.437	1.982	1.8	1.227
T cf. *reepeni*	Ellipsoid	1.353	1.624	1.49	1.09
T. laeviceps (Java)	Ellipsoid	1.424	1.843	1.68	1.277
T. cf. *pagdeniformis*	Cuboid	0.98-1.09	2.152	1.614	1.333
T. cf. *ruficornis*	Ovaloid	1.264	1.506	1.365	1.103
T. aff. *sarawakensis*	Ovaloid	1.2652	1.81	1.64	1.103
T. fuscobalteata	Ovaloid	1.308	1.241	1.04	1.194
**T, carbonaria* NSW	Ovaloid	1.276	1.93	-	-
**T. carbonaria* Qld	Ovaloid	1.346	2.08	-	-
**Austroplebeia* sp.	Ovaloid	1.25	1.745	1.418	1.057

For the Indo-Malayan region, the *Tetragonula* Queen cells and 1.5 to 2.275 times larger than the larval cells and 1.365 to 1.8 times larger than the pupal cells., Except for *T. fuscobalteata,* where the Queen cell width/Larval cell width ratio is 1.241, and pupal cell width is 1.04.

For the Australasian region, the queen cell is very conspicuous at the retreating edge periphery and are 1.9 to 2.1 times larger than the other brood cells. I cannot detect any difference in cell width between the larval and pupal cells at the current resolution in the *T. carbonaria* image bank. For comparisons, I include data on an *Austroplebeia* sp. where the Queen cell is 1.5 to 1.75 times larger than the other brood cells.

Figure 145 Austroplebeia sp.

[14] Q-cell h/w = Queen Cell height/width; Qw/Lw = Queen cell width/Larval cell width; Qw/Pw = Queen cell width/Pupal cell width; Pw/Lw = Pupal cell width/Larval cell width.

11.6 Image Gallery of *Tetragonula* Queens and some.

Images are sorted by species alphabetically.

(aff. *T. bengalensis or sp.*) - Thungben Yungthen. Nagaland,

T. biroi - Mindanao

T. biroi - Sherwin Baja, Bicol

T. nr. *biroi* – Biliran, Leyte

T. cf *biroi*, Gerder, SE Sulawesi

T. nr. *Biroi* - Jr Belmonte, Luzon

T. carbonaria - Dan Smailes, Sydney

T. carbonaria - Dan Smailes, Sydney

T. carbonaria virgin queen - Giorgio Venturieri

T. clypearis - Sian Lawson, Mackay

T. clypearis, Nick Powell, Cape York

T. cf. *clypearis* - Gogo Manug, Bukidnon

T. cf *dapitanensis* – AHJ, Davao

T. cf. *dapitanensis* - Elyah Albara, Zamboanga City

T. drescheri - Teguh, Banyumas, Java,

T. fuscobalteata young queen – Rusphin, Ranau, Sabah

T. fuscobalteata – Hisyam Ismail, KKB, Selangor

T. fuscobalteata - Alrahman P. Pasi, Sumatra

T. melanocephala 1 day old - Ahmad Sapruddin, Borneo

T. melanocephala Ahmad Sapruddin, Banjarbaru,

T. melina - Man Tool - My Iqra – SFV, Selangor.

T.* cf. *mellipes La Hisa – found in a termites' nest in Merauke.

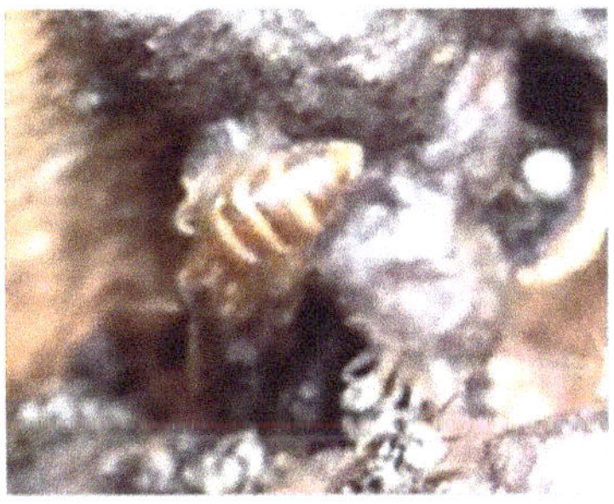

T. aff. minangkabau Kamarul Nasir, Kelantan

T. cf. *pagdeni or sp* Albi Banjarmasin

T. reepeni – Alrahman P. Pasi, Aceh, N. Sumatra.

*T.*cf. *ruficornis* or sp. - Subrahmanya, Karnataka

T. aff. *sarawakensis* or sp. - Bode Shuya, Nagaland ,

T. aff. *sarawakensis* or sp - Stephen Peterson, Cambodia.

T. cf *drescheri* - Omah Klanceng, Purworejo, Java

Tetragonilla atripes
Zul Ini, Kelantan

Tetragonula sp. - Gogo Manug, Mindanao

Tetragonula sp. - Trình Nguyễn, Vietnam

T. sapiens – Melvin Guerrero, Isabel. Leyte

T. sapiens – Nora A. Peña, Iloilo, Panay

T. sapiens gyne - Suardi Saide, Sulawesi

T. Sapiens - Nick Powell, *Brisbane, Qld.*

T. sapiens - Zac Peterson, Cairns, Queensland

T. sapiens queen - Suardi Saide, Sulawesi

T. cf. *testaceitarsis* - Gogo Manug, Bukidnon, Mindano

T. cf. *testaceitarsis* - Gogo Manug, Bukidnon, Mindano

T. testaceitarsis - Abu Hassan Tg Ipoh, N9

***T.* cf. *testaceitarsis or* sp. -** Rahmat, Riau, Sumatra

***T.* cf. *testaceitarsis or* sp. -** Rahmat, Riau, Sumatra

***T.* cf. *testaceitarsis or* sp. -** Albi Banjarmasin, S.Kalimantan

Austroplebeia australis - Bob Luttrell, Tiaro. Queensland.

Lophotrigona canifrons - Muhammad Asraf, Johor

Lisotrigona aff. cacciae Agus Suhartanto, E. Borneo

Austroplebeia australis , Dan Smailes, Sydney, NSW.

Wallacetrigona incisa - Hassan Asri, Sulawesi

Lisotrigona furva - Pichet Yiangyong, Bangkok, Thai.

Chapter 12

12.0 Opercula and cell surface textures

Opercula, plural of the operculum, refers to the cap of the Brood cell. The caps vary with the different shapes of the brood cells.

12.1 Oblate Spheroid

The first shape we look at is the Oblate spheroid. An example is one of the *T. minangkabau* brood cells on the left. The one on the right is an example of *Austroplebeia magna*, where the average h/m = 1.01.

Figure 146 Oblate Spheroid cell textures

Just like the Moon's surface, it is not all that smooth when zoomed in. The crest of the cap has tiny hillocks. We can make out the seam of the cap brim made with the cell body with closer inspection. The cap to cell body seam is not well rounded as one might think, and they are quite irregular. The irregularity is not a lid per se but a build-up of wax and propolis particles. The cell textures range from crumbly to mottled with smeared colouration.

12.2 Prolate spheroid

Similarly, a closeup of the Prolate spheroids shows a crumbly surface of the cells, a *Tetragonula* sp. in Mindanao, Philippines. The crumbly texture is pronounced on macro photograph images, and it reflects that the cells are not oviposited whole like a bird's egg but built-up particle by particle of wax mostly.

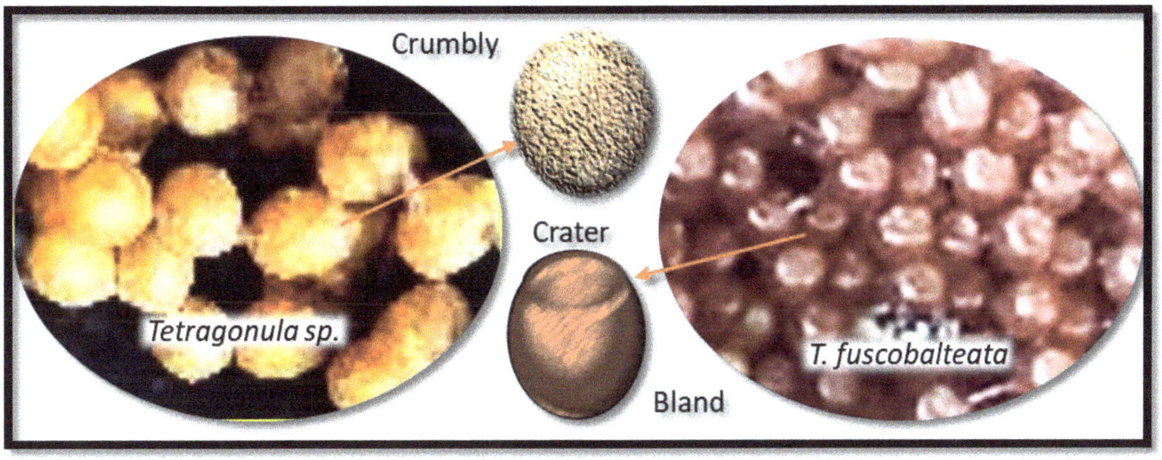

Figure 147 Prolate Spheroid cell textures

In some instances, the caps appear depressed in the new brood, and the caps bloat up as the larva grows. This is observed in *T. fuscobalteata* in Borneo. The cell texture is relatively bland, with no markings besides a crater-like depression.

12.3 Ovaloid

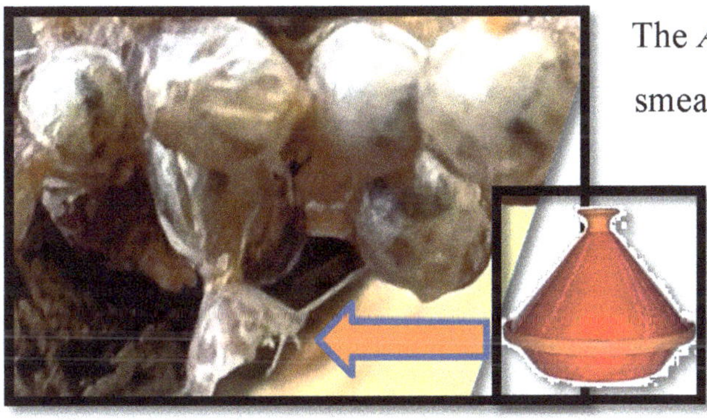

The *Austroplebeia* sp. cell texture in Australia has smeared colouration. The opercula remind me of the Moroccan Tagine clay pot cover, and they are top of the cover sometimes serve as connectives between cells. This operculum is also translucent, and one can view the eyes of the imago within.

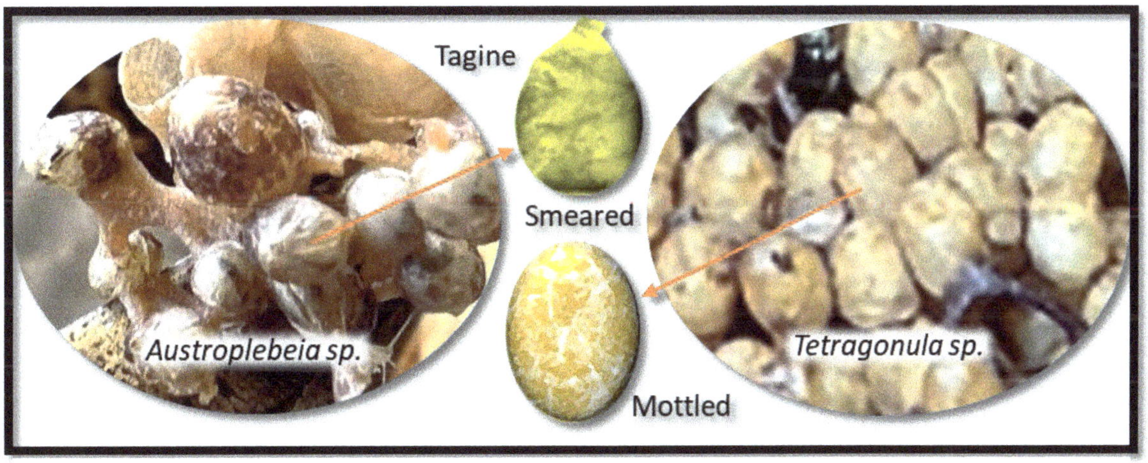

Figure 148 Ovaloid cell textures

The *Tetragonula* sp. *T. sapiens* or *T. testaceitarsis* in the Visayas has distinct mottling. The opercula here may have some wrinkles.

12.4 Ellipsoid

Figure 149 Ellipsoid cell textures

Ellipsoidal brood cell textures for *T. sirindhornae* in Peninsula Malaysia appear wrinkled, and the opercula are fused with no marked distinction of the seams. For *T. laeviceps* in Borneo, the cell texture is bland and rippleless.

12.5 Cylindroid

This shape is usually associated with semi-combs or full wafered or stacked combs. Those in *T. davenporti* in Australia appeared mottled in macro photograph images, and the opercula are domed. For *T.* nr. *biroi in Sulawesi,* the cell textures are bland, and no colouration smears or mottling.

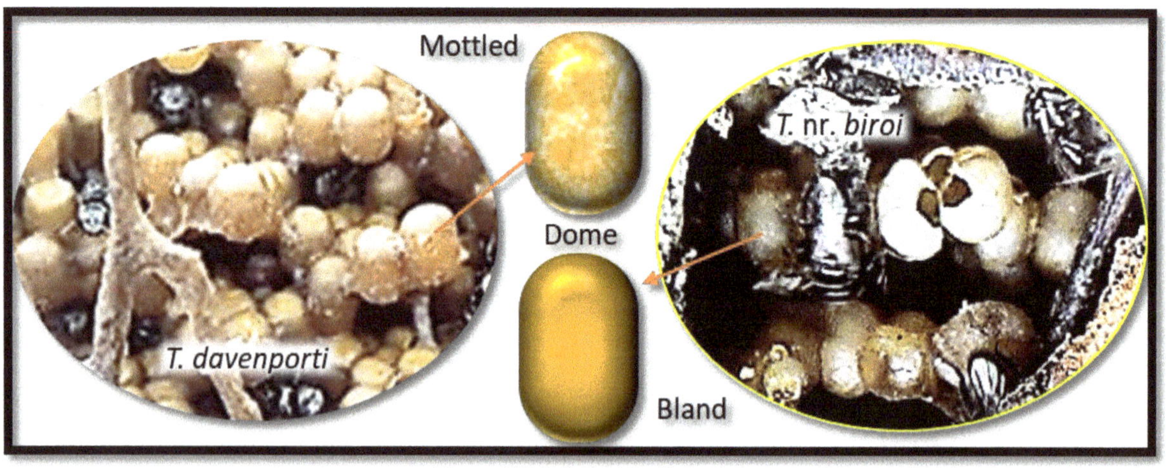

Figure 150 Cylindroid cell textures

12.6 Translucent opercula

Figure 151 Translucent opercula of cocoons in Lophotrigona and Lisotrigona.

Translucent opercula of cocoons in *Lophotrigona* and *Lisotrigona*. The dark eyes of the imagos are visible through the opercula. This translucence sometimes allows such visibility in *Austroplebeia*, and I have not observed this phenomenon explicitly in *Tetragonula*. It may occur in the pale cream pupal cells but is not distinguishable in the darker coloured cocoons.

End of Part 2.

Part 3

History and Chronology of Indo-Malayan and Australasian clades of Stingless Bees (Apidae: Meliponini) Taxa and Trends - Prehistoric to present day.

Tabulations: History of Tetragonula taxonomical combinations.

Table 3 History and chronology of Tetragonula Synonymy

Taxa	Author	Year	Synonym or alternative combination
	Moure	1961	***Tetragonilla***
	Moure	1961	***Tetragonula***
	Shanas	2019	***Flavotetragonula, subgen. n.***
Flavotetragonula calophyllae	Shanas and Faseeh	2019	*Tetragonula (Flavotetragonula) calophyllae*, n. sp.
Tetragonilla atripes	Smith	1857	*Trigona atripes*
Tetragonilla collina	Smith	1857	*Trigona collina*
Tetragonilla fuscibasis	Cockerell	1920	*Trigona fuscibasis*
Tetragonilla rufibasalis	Cockerell	1918	*Trigona rufibasalis*
Tetragonula malaipanae	Engel, Michener & Boontop	2017	*Tetragonula (Tetragonula) malaipanae*
Tetragonula basimaculata	Bingham	1903	*Melipona basimaculata*
Tetragonula bengalensis	Cameron	1897	*Trigona bengalensis*
Tetragonula biroi	Friese	1898	*Trigona biroi*
Tetragonula carbonaria	Smith	1854	*Trigona carbonaria*
Tetragonula clypearis	Friese	1909	*Trigona laeviceps* var. *clypearis*
Tetragonula dapitanensis	Cockerell	1925	*Trigona dapitanensis*
Tetragonula davenporti	Franck	2004	*Trigona davenporti*
Tetragonula devicta	Cockerell	1921	*Trigona (Tetragonula) devicta*; *Meliponorytes devict*
Tetragonula drescheri	Schwarz	1939	*Trigona (Tetragona) sarawakensis var drescheri*
Tetragonula fuscobalteata	Cameron	1908	*Trigona fuscobalteata*
Tetragonula geissleri	Cockerell	1918	*Trigona geissleri*
Tetragonula gressitti	Sakagami	1978	*Trigona (Tetragonula) gressitti*
Tetragonula hirashimai	Sakagami	1978	*Trigona (Tetragonula) hirashimai*
Tetragonula hockingsi	Cockerell	1929	*Trigona carbonaria hockingsi* n. subsp.
Tetragonula iridipennis	Smith	1854	*Trigona iridipennis*
Tetragonula kyrdemkulaiensis	Viraktamath and Rojeet	2020	*Tetragonula kyrdemkulaiensis, sp. n.*
Tetragonula laeviceps	Smith	1857	*Trigona laeviceps*
Tetragonula melanocephala	Gribodo	1893	*Trigona melanocephala*
Tetragonula melanocephala	Cameron	1908	*Trigona testaceinerva*
Tetragonula melina	Gribodo	1893	*Trigona melina*
Tetragonula mellipes	Friese	1898	*Trigona mellipes*
Tetragonula minangkabau	Sakagami & Inoue	1985	*Trigona (Tetragonula) minangkabau*
Tetragonula minor	Sakagami	1978	*Trigona (Tetragonula) minor*
Tetragonula pagdeni	Schwarz	1939	*Trigona (Tetragona) fusco-balteata var. pagdeni*
Tetragonula pagdeniformis	Sakagami	1978	*Trigona (Tetragonula) pagdeniformis*
Tetragonula penangensis	Cockerell	1919	*Trigona penangensis*
Tetragonula perlucipinnae	Faseeh and Shanas	2019	***Tetragonula (Tetragonula) perlucipinnae, n. sp.***
Tetragonula praeterita	Walker	1860	*Trigona præterita*
Tetragonula reepeni	Friese	1918	*Trigona reepeni*
Tetragonula ruficornis	Smith in Horne & Smith,	1870	*Melipona ruficornis*
Tetragonula sapiens	Cockerell	1911	*Trigona sapiens*
Tetragonula sarawakensis	Schwarz	1937	*Trigona sarawakensis*
Tetragonula sirindhornae	Michener & Boongird	2004	*Trigona (Tetragonula) sirindhornae*
Tetragonula srikantanathi	Viraktamath	2020	*Tetragonula srikantanathi, sp. n.*
Tetragonula testaceitarsis	Cameron	1901	*Trigona testaceitarsis*
Tetragonula travancorica	Faseeh and Shanas	2019	***Tetragonula (Tetragonula) travancorica, n. sp.***
Tetragonula valdezi	Cockerell	1918	*Trigona valdezi*
Tetragonula zucchii	Sakagami	1978	*Trigona (Tetragonula) zucchii*

Prelude to Part 3: This section may appear as a stand-alone article but slot in here as a subordinate part of brood cells and form classification. It depicts the trends and history of naming stingless bees and their synonymy and why they are associated with the classifications in Parts 1 & 2. It forms a prequel to the image slides of relevant regional species in Part 4.

This 3rd part looks at the origin of the names from the first occurrence of the initial Stingless Bee Genus and how the names change and evolve chronologically. The names of the earliest Meliponine fossils are looked into. A brief biography of the name authors and taxonomists is given and the trend they were inclined to emulate.

The new names and their validity are examined, and relevant taxonomists suggest the synonyms. The names proposed in parts 1 & 2 were put there because they had brood images associated with them, and this part provides scrutiny of the possible credibility or mistakes of those names.

Over the years, we have seen taxa evolve from naming the bees according to the physical attributes from prehistoric periods as old as the Cretaceous (about 70 Mya) to present-day discoveries of new species as recent as Dec 2017. Even the origin that the fossils occurred, e.g. Dominican Republic amber - *Proplebeia dominicana* were used, to Patronyms, e.g. *Tetragonula davenporti Franck 2004* and Matronyms, e.g. *Tetragonula malaipanae* n.sp. 2017 & and even a Theonym in India e.g *Lisotrigona revanai* n. sp. 2017. Before this Patronymic trend, the bees were named by their colour attributes like *canifrons, atripes*, or *itama* and *carbonaria* until they practically ran out of colour descriptions. Then they named the bees where they found them like *minangkabau*, *sarawakensis*, *sandacana* or *palavanica*.
This study also provides an illustrated key to identifying Indo-Malayan stingless bees. It is designed to be especially useful as a teaching aid for beekeepers, students and hobbyists. The Indo-Malayan stingless bee genera key is especially critical to classify those dependent on dipterocarp trees. The conservation of stingless bees and dipterocarp trees relies on a pictorial Key. The incorporated 'Illustrated Key' too has evolved into a new key incorporating the Australasian & Papuasia clades with a revamp of placements of genera and additions of new subgenus.

Chapter 13

13.0 History and Chronology

13.0.1 Introduction

The stingless bees name from as long ago as 1806 (by Johann Karl Wilhelm Illiger, a German entomologist and zoologist) (Wikipedia, Johann Illiger, 2014) as in *Melipona* (Illiger 1806) and also 1807 (by a Swiss Physician, Louis Jurine) (Wikipedia, Louis Jurine, 2014) with *Trigona* (Jurine 1807) are a testament to the Linnean Taxonomy (Biological Classification of Living things) of 1735 with the naming of animals in the 10th edition in 1758. Biological classification, or scientific classification in biology, is a method of scientific taxonomy used to group and categorize organisms into groups in a hierarchical way, such as genus or species.

Portrait 1 Luis Jurine (Wikipedia, Louis Jurine, 2014)

Figure 152 Dominican Amber Specimen: HYMENOPTERA Apidae: Proplebeia dominicana (Stingless Bee) Possibly 17 mya (typical of Dominican Amber)

13.1 Meliponini

Stingless bees, sometimes called **stingless honey bees** or simply **meliponines**, are a large group of bees (approximately 600 species), comprising the tribe **Meliponini** (or subtribe Meliponina according to some authors). They belong in the family Apidae and are closely related to common honey bees, carpenter, orchid bees and bumblebees. ***Melipona*** Illiger, K. (1806) was then adopted, as per Moure's Bee catalogue: Type species: *Apis favosa* Fabricius, 1798 by Subsequent designation, Latreille, PA (1810).

The same nomenclature codes govern even names of extinct Meliponine species, e.g., the extinct genus *Proplebeia* (Figure 1) are recognized four species, three of them from the amber of Dominican Republic, hence the name *Proplebeia dominicana.*

To better comprehend the taxon used in naming stingless bees, it is imperative to compile a list of the derivative nomenclature, including prefixes, suffixes and root words that comprise the naming structure. The reader will recognize geographical locations and names of naturalists involved in the study of bees in general because many species were named after these researchers and locations of where they were collected or observed. Other terminology, basically Latin, describes the anatomy characteristics and/or colour, and this allows easier identification by simply identifying location observed/recorded, anatomy description and colour.

These seemingly exhaustive definitions are not certified specifics of the taxonomical definitions but merely a listing of possible lingual explanations of the reasons for the naming the way the nomenclature was structured. It is intended for the layman beekeeper and enthusiast to recognize the bees' anatomical description and characteristics from their indicative names. The names of the naturalists, researchers and collectors are also honoured accordingly in the validated taxa.

Meliponini = Translingual – **Etymology:** - *Melipona* (Latin Meli (honey) + pona (latin "I make")) + *-ini* (see suffix –ini) - Meliponini - A taxonomic **tribe** within the **subfamily Apinae**

-ini = Translingual – **Etymology: New Latin**, the nominative plural form of **Latin *-inus*** - **Suffix'**–ini' = **(taxonomy) Used to form names of tribes of animals**. The tribe Meliponini was established by Lepeletier in 1844 (in Michener 2000)

13.1.1 Trigona

Trigona is used as a genus name, suffix, and prefix in the many names. It puts a need to define the term and explore its significance. The Fairfax Medical dictionary defines trigona as follows: tri·go·na (trī-gō'nă) = Plural of trigonum. Definition of *TRIGONUM*: a triangular anatomical part. Triangular Anatomical features of the stingless bee (see Figure 2 & Figure 3)

1) **cephalic triangle** on cranium formed by lines connecting the metopion, the pogonion, and the occipital point.
2) **clypeal triangle,** the fronto-clypeal suture forms a 'supra-clypeal **triangle.**'
3) **digastric triangle** submandibular **triangle**.

4) **facial triangle is** the area that is raised and extends from

between the toruli to the posterior margin of the clypeus

5) **occipital triangle**. the opening where organs of head and body are connected

6) **ocellar triangle** typically three bead-like lenses arranged in a **triangle**

7) **propodeal triangle** of propodeum

8) **tibial triangle** triangularly clavate hind tibia

9) **tarsal triangle** triangular shape of tarsus

Figure 153 Propodeum of Geniotrigon thoracica propodeum with shiny bare spot. photo courtesy of Deborah Smith

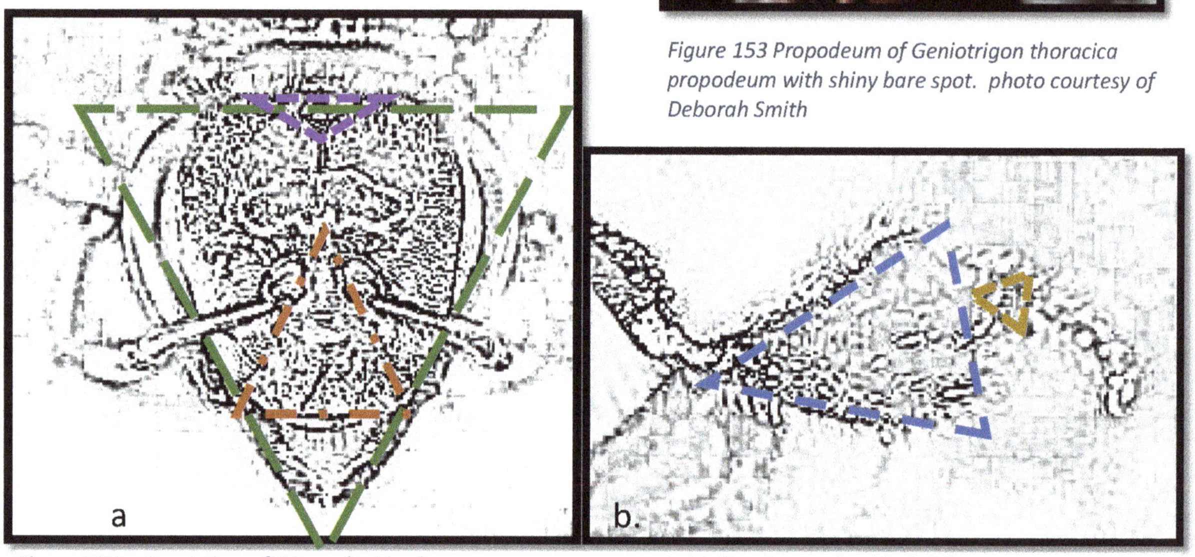

Figure 154 a. Front view of Triangular Head; *b. Meliponine worker Hind tibia*

13.2 Prehistoric Meliponines

Table 4 Prehistoric Meliponines sorted by amber age

Item#	Fossilized Genus	Author	Year	Fossil Discovered	Age (Ma)
1.0	***Trigona (Trigona)***	Jurine	1807	Quaternary of Myanmar (1 collection), Miocene of the Dominican Republic (1), Italy (1), Eocene of the Russian Federation (2)	
1.1	*Trigona (Trigona) erythra*	Schletterer	1891	*Melipona (Trigona)* ***erythra***[2]	
2.0	***Trigonisca***	Moure	1950	Columbian Copal	1.6 – 0.0

2.1	*T. sp*	Engel	2001	Columbian Copal	1.6 – 0.0
3.0	***Ptilotrigona***	Moure	1951	Columbian Copal	
3.1	*Ptilotrigona lurida*	Smith	1854		
4.0	***Trigona (Tetragona)***	Lepeletier and Serville	1828	Burmese copal - Pleistocene	0.012 to 0.0005
4.1	*Trigona (Tetragona) iridipennis*	Smith	1854	Burmese copal	0.012 to 0.0005
5.0	***Meliponula***	Cockerell	1934	East African copal - Pleistocene	0.012 to 0.0
5.1	*Meliponula (Axestotrigona) ferruginea*[2]	Lepeletier	1841	*Melipona (Trigona) ferruginea* Lepeletier, 1841	
6.0	***Liotrigona***	Moure	1961	Pleistocene	0.012 to 0.0
6.1	*Liotrigona vetula*	Moure and Camargo	1978	Copal	0.012 to 0.0
7.0	***Tetragonula***	Moure	1961	Pleistocene	
7.1	*Trigona (Tetragonula) devicta*	Kerr & Maule	1964	Earlier ***T. iridipennis***, (Smith 1854)[3] (Cockerell 1921) – Burmese Copal	0.012 to 0.0
8.0	***Meliponorytes***	Tosi	1896	Miocene	20
8.1	*Meliponorytes devictus*	Cockerell	1921	Sicilian amber	20
8.2	*Meliponorytes succini*	Tosi	1896	Sicilian amber	20
8.3	*Meliponorytes sicula*	Tosi	1896	Miocene - Sicilian Amber	20
9.0	***Proplebeia***	Michener	1982	Miocene	20.43 to 13.65
9.1	*Proplebeia abdita*	Greco and Engel	2011	Dominican amber	17
9.2	*Proplebeia dominicana*	Wille and Chandler	1964	Dominican amber	17
9.3	*Proplebeia tantilla*	Camargo	2000	Dominican amber	17
9.4	*Proplebeia vetusta*	Camargo	2000	Dominican amber	17
10.0	***Nogueirapis***	Moure	1953	Miocene of Mexico	23.03 to 15.97
10.1	*Nogueirapis silacea*[1]	Wille	1959	Miocene of Mexico	23.03 to 15.97
11.0	***Kelneriapis***	Sakagami	1978	*Syn. Kelnermelia* Moure 1978	*no occurrences*
11.1	*Trigona (Hypotrigona) dominicana*	Wille and Chandler	1964	N Comb. *Proplebeia dominicana*	17
11.2	*Trigona (Hypotrigona) eocenica*	Kelner-Pillault	1970	Baltic Amber	45.1

11.3	*Trigona (Hypotrigona) gribodoi*	Magretti	1884	*no occurrences*	*no occurrences*
12.0	***Liotrigonopsis***	Engel	2001	Baltic Amber	45.1
12.1	*Liotrigonopsis rozeni*	Engel	2001	Baltic Amber	45.1
13.0	***Exebotrigona***	Engel and Michener	2013	Fushun amber (Eocene of China)	55.8 to 48.6
13.1	*Exebotrigona velteni*	Engel and Michener		n. sp.	55.8 to 48.6
14.0	***Cretotrigona***	Engel	2000	New Jersey amber – Late Cretaceous	70.6 to 66.043
14.1	*Cretotrigona prisca*	Michener and Grimaldi	1988	New Jersey amber	70.6 to 66.043

(1) Camargo & Pedro (2007) have placed this species in *Proplebeia*. However, "When *Nogueirapis silacea*, as originally described in the literature (Wille & Chandler 1964), and the very clear hind leg drawing among the four presented, were reviewed by Camargo et al. (2002), those authors changed the genus name to *Proplebeia*. Engel (2004) overturned this change in the genus, finding that the holotype is not *Proplebeia*, but I (D. W. Roubik) found that at least one paratype is certainly not *Nogueirapis*." (Roubik 2012)

(2) Originally recorded as Meliponula erythra (Schletterer, 1891), a subjective junior synonym of M. ferruginea (Eardley, 2004)

(3) Originally described as Meliponorytes devictus Cockerell 1921 (Sakagami, 1978; Moure & Camargo, 1978)

13.3 Oldest Fossil Stingless Bee

Cretotrigona prisca (Michener and Grimaldi), new combination *Trigona* (*Trigona*) *prisca* Michener and Grimaldi, 1988. The graphics below are a digital rendering based on photomicrographs of holotype worker of *Cretotrigona prisca* (AMNH C88720). And descriptions in "A New Interpretation of the Oldest Fossil Bee (Hymenoptera: Apidae)" Michael S. Engel 2000.

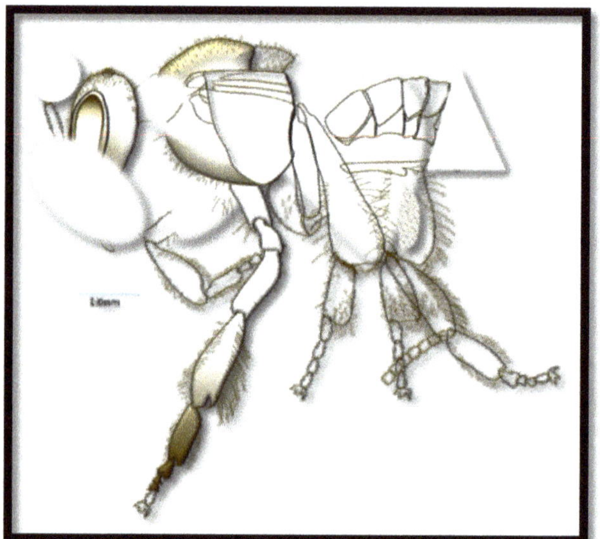

Figure 155 *An impression of how the oldest stingless bee Cretotrigona prisca, would have looked 65 mya.*

13.4 Ichnospecies

The International Code of Zoological Nomenclature defines an ichnotaxon (plural ichnotaxa) as "a taxon based on the fossilized work of an organism", that is, the non-human equivalent of an artefact. The International Code of Zoological Nomenclature defines an ichnotaxon (plural ichnotaxa) as "a taxon based on the fossilized work of an organism", that is, the non-human equivalent of an artefact-. *Ichnotaxa* comes from the Greek ἴχνος, *ichnos* meaning *track* ταξις, and *taxis* meaning *ordering; arrange or set in order.*

Ichnotaxa are names used to identify and distinguish morphologically distinctive ichnofossils, more commonly known as trace fossils. They are assigned genus and species ranks by ichnologists, much like organisms in Linnaean taxonomy. These are known as **ichnogenera** and **ichnospecies**, respectively. "Ichnogenus" and "ichnospecies" are commonly abbreviated as "igen." and "isp.". The binomial names of ichnospecies and their genera should be written in italics.

Most researchers classify trace fossils only as far as the ichnogenus rank, based upon trace fossils that resemble each other in morphology but have subtle differences. Some authors have constructed detailed hierarchies up to ichnosuperclass, recognizing such fine detail as identifying ichnosuperorder and ichnoinfraclass, but such attempts are controversial.

Figure 156 top left, Heterotrigona itama dual entrance tube; top right, a trace fossil of dual entrance tube on silicified shale; mid-left, Meliponine robbing nectar; mid-right, Meliponine pinching for phloem at leaf shoots.; bottom left, leaf damage result of shoot pinching; bottom right, a trace fossil of damaged leaf possible evidence of shoot pinching.

13.5 Chronology of Meliponine Taxa (19th – 20th Century)

Table 5 Chronology of Modern Nomenclature in Meliponine Taxonomy (19th – 20th Century)

TEXT COLOR CODE

1. **Black = Tribe**
2. **Grey = Nomen dubium / Nodum**
3. **Bright red = Genera**
4. **Orange = Subgenera**
5. **Brown = Patronyms & Matronyms**
6. **Purple = Peoples & Places**
7. **Red = Anatomy & color attributes**
8. **Blue = Flora and Fauna**
9. **Green = General epithets**
10. L. purple = * Extinct species

Year	Author	Taxa	Synonym or alt. comb.
1806	Illiger	***Melipona***	
1807	Jurine	***Trigona***	
1828	Lepeletier & Serville	***Tetragona***	
1836	Lepeletier	*Melipona vidua*	*Tetrigona vidua*
1844	Lepeletier	Meliponini	n. tribe
1854	Smith	*Trigona carbonaria*	*Tetragonula carbonaria*
1854	Smith	*Trigona iridipennis*	*Tetragonula iridipennis*
1857	Smith	*Trigona collina*	*Tetragonilla collina*
1857	Smith	*Trigona apicalis*	*Tetrigona apicalis*
1857	Smith	*Trigona atripes*	*Tetragonilla atripes*
1857	Smith	*Trigona canifrons*	*Lophotrigona canifrons*
1857	Smith	*Trigona fimbriata*	*Homotrigona fimbriata*
1857	Smith	*Trigona laeviceps*	*Tetragonula laeviceps*
1857	Smith	*Trigona nitidiventris*	*Lepidotrigona nitidiventris*
1857	Smith	*Trigona thoracica*	*Geniotrigona thoracica*
1857	Smith	*Trigona ventralis*	*Lepidotrigona ventralis*
1860	Walker	*Trigona prœterita*	*Tetragonula praeterita*
1865	Smith	*Trigona atricornis*	*Platytrigona atricornis*
1865	Smith	*Trigona planifrons*	*Platytrigona planifrons*
1870	Smith in Horne & Smith,	*Melipona ruficornis*	*Tetragonula ruficornis*
1878	Smith	*Trigona terminata*	*Lepidotrigona terminata*
1891	Gribodo	*Trigona javanica*	*Lepidotrigona javanica*
1893	Gribodo	*Trigona melanocephala*	*Tetragonula melanocephala*
1893	Gribodo	*Trigona melina*	*Tetragonula melina*
1896	Dalla Torre	*Melipona pulla*	*Nomen nudum.*
1897	Bingham	*Trigona ruficornis* Smith	*Melipona smithii*
1897	Bingham	*Melipona lutea*	*Homotrigona lutea*
1898	Friese	*Trigona biroi*	*Tetragonula biroi*
1898	Friese	*Trigona australis*	*Austroplebeia australis*

1898	Mocsáry dalam Friese	*Trigona cincta* subsp. *percinta* nov.	*Austroplebeia cincta*
1898	Friese	*Trigona mellipes*	*Tetragonula mellipes*
1900	Friese	*Trigona latipes*	*Lepidotrigona latipēs*
1901	Friese	*Trigona keyensis*	*Platytrigona keyensis*
1901	Cameron	*Trigona testaceitarsis*	*Tetragonula testaceitarsis*
1902	Cameron	*Trigona erythrogastra*	*Heterotrigona erythrogastra*
1902	Cameron	*Trigona flavistigma*	*Homotrigona fimbriata*
1902	Cameron	*Trigona lacteifasciata*	*Geniotrigona lacteifasciata*
1902	Cameron	*Trigona latebalteata*	*Lepidotrigona latebalteata*
1903	Bingham	*Melipona basimaculata*	Nomen dubium
1903	Bingham	*Melipona castanea*	*Homotrigona fimbriata*
1905	Cockerell	*Trigona essingtoni*	*Austroplebeia essingtoni*
1907	Nurse	*Melipona cacciae*	*Lisotrigona cacciae*
1908	Cameron	*Trigona erythrostoma*	*Tetragonula fuscobalteata*
1908	Cameron	*Trigona fulvopilosella*	*Lepidotrigona nitidiventris*
1908	Cameron	*Trigona fuscobalteata*	*Tetragonula fuscobalteata*
1908	Cameron	*Trigona pallidistigma*	*Tetragonula fuscobalteata*
1908	Cameron	*Trigona testaceinerva*	*Tetragonula melanocephala*
1909	Friese	*Trigona laeviceps* var. *clypearis*	*Tetragonula clypearis*
1909	Friese	*Trigona flaviventris*	*Platytrigona flaviventris*
1909	Friese	*Trigona genalis*	*Papuatrigona genalis*
1909	Friese	*Trigona luteiventris*	*Heterotrigona erythrogastra*
1909	Friese	*Trigona versicolor*	*Homotrigoma fimbriata*
1909	Friese	*Trigona anamitica*	*Homotrigona anamitica*
1910	Cockerell	*Trigona cassiae*	*Austroplebeia cassiae*
1911	Cockerell	*Trigona sapiens*	*Tetragonula sapiens*
1912	Cockerell	*Trigona angophorae*	*Tetragonula carbonaria*
1913	Strand	*Trigona ventralis hoozana*	*Lepidotrigona hoozana*
1915	Cockerell	*Trigona palavanica*	*Lepidotrigona palavanica*
1918	Cockerell	*Trigona geissleri*	*Tetragonula geissleri*
1918	Friese	*Trigona reepeni*	*Tetragonula reepeni*
1918	Cockerell	*Trigona valdezi*	*Tetragonula valdezi*
1918	Cockerell	*Trigona ambusta*	*Geniotrigona thoracica*
1918	Cockerell	*Trigona itama*	*Heterotrigona itama*
1918	Cockerell	*Trigona melanotricha*	*Homotrigona anamitica*
1918	Cockerell	*Trigona pallidicincta*	Incerta sedis / Nomen dubium
1918	Cockerell	*Trigona rufibasalis*	*Tetragonilla rufibasalis*
1918	Cockerell	*Trigona busara*	*Lophotrigona canifrons*
1919	Cockerell	*Trigona bakeri*	*Heterotrigona bakeri*
1919	Cockerell	*Trigona penangensis*	*Tetragonula penangensis*

1919	Cockerell	*Trigona sandacana*	*Heterotrigona erythrogastra*
1919	Cockerell	*Trigona breviceps*	*Heterotrigona itama*
1919	Cockerell	*Trigona fulvomarginata*	*Lepidotrigona terminata*
1919	Cockerell	*Trigona hæmatoptera*	*Odontotrigona haematoptera*
1919	Cockerell	*Trigona atomella*	*Tetragonula fuscobalteata*
1919	Cockerell	*Trigona confusella*	*Tetragonula geissleri*
1920	Cockerell	*Trigona fuscibasis*	*Tetragonilla fuscibasis*
1920	Cockerell	*Trigona scintillans*	*Lisotrigona cacciae* (Nurse, 1907)
1920	Cockerell	*Trigona trochanterica*	*Lepidotrigona trochanterica*
1921	Cockerell	*Meliponorytes devictus* *	*Trigona (Tetragonula) devicta**
1921	Cockerell	*Trigona (Tetragonula) devicta**	*Tetragonula devicta* *
1925	Friese	*Trigona pfeifferi*	*Tetragonula fuscobalteata*
1925	Cockerell	*Trigona dapitanensis*	*Tetragonula dapitanensis*
1926	Cockerell	*Trigona cambodiensis*	*Tetragonilla collina*
1927	Cockerell	*Trigona apicalis peninsularis* subsp.n.	*Tetrigona peninsularis*
1929	Cockerell	*Trigona aliceae*	*Homotrigona aliceae*
1929	Cockerell	*Trigona carbonaria hockingsi* n. subsp.	*Tetragonula hockingsi*
1929	Dover	*Trigona kusutkana*	*Homotrigona aliceae*
1929	Cockerell	*Trigona lamingtonia*	*Platytrigona lamingtonia*
1929	Cockerell	*Trigona wybenica*	*Tetragonula clypearis*
1929	Cockerell	*Trigona ferrea*	*Homotrigona lutea*
1929	Cockerell	*Trigona flavibasis*	*Lepidotrigona flavibasis*
1929	Cockerell	*Trigona hemileuca*	*Tetrigona apicalis*
1929	Cockerell	*Trigona melanoleuca*	*Tetrigona melanoleuca*
1929	Cockerell	*Trigona cincta percincta*	*Austroplebeia percincta*
1929	Cockerell	*Trigona arcifera*	*Lepidotrigona arcifera*
1930	Rayment	*Trigona cockerelli*	*Austroplebeia cockerelli*
1932	Rayment	*Trigona cockerelli ornata* subsp.	*Trigona (Heterotrigona) davenporti*
1932	Rayment	*Trigona symei*	*Austroplebeia symei*
1932	Rayment	*Trigona websteri*	*Austroplebeia websteri*
1932	Rayment	*Trigona cockerelli ornata*	*Austroplebeia ornata*
1933	Friese	*Trigona borneënsis*	*Geniotrigona lacteifasciata*
1933	Friese	*Trigona pygmaea*	*Tetragonula fuscobalteata*
1933	Friese	*Trigona sericea*	*Tetrigona apicalis*
1934	Cockerell	***Hypotrigona***	
1937	Schwarz	*Trigona apicalis var. binghami*	*Tetrigona binghami*
1937	Schwarz	*Trigona hobbyi*	*Platytrigona hobbyi*
1937	Schwarz	*Trigona moorei*	*Sundatrigona moorei*

1937	Schwarz	*Trigona haemanoptera var. dulitae*	*Odontotrigona haematoptera*
1937	Schwarz	*Trigona sarawakensis*	*Tetragonula sarawakensis*
1938	Schwarz	***Plebeia***	
1939	Schwarz	***Heterotrigona***	
1939	Schwarz	***Lepidotrigona***	
1939	Schwarz	*Trigona (Tetragona) sarawakensis var drescheri*	*Tetragonula drescheri*
1939	Schwarz	*Trigona (Hypotrigona) pendleburyi var. klossi*	*Pariotrigona klossi*
1939	Schwarz	*Trigona (Tetragona) fusco-balteata var. pagdeni*	*Tetragonula pagdeni*
1939	Schwarz	*Trigona (Hypotrigona) pendleburyi*	*Pariotrigona pendleburyi*
1939	Schwarz	*Trigona (Lepidotrigona) ventralis* var. *doipaensis*	*Lepidotrigona doipaensis*
1947	Pugh	*Melipona khasiana*	*Nomen Nudum*
1947	Pugh	*Melipona terestris*	*Nomen Nudum*
1947	Pugh	*Melipona tunneli*	*Nomen dubium*
1959	Sakagami	*Trigona (Tetragona) matsumurai*	*Sundatrigona moorei*
1961	Moure	***Austroplebeia***	
1961	Moure	***Geniotrigona***	
1961	Moure	***Homotrigona***	
1961	Moure	***Lisotrigona***	
1961	Moure	***Lophotrigona***	
1961	Moure	***Odontotrigona***	
1961	Moure	***Pariotrigona***	
1961	Moure	***Platytrigona***	
1961	Moure	***Tetragonilla***	
1961	Moure	***Tetragonula***	
1961	Moure	***Tetrigona***	
1975	Sakagami et Moure	***Trigonella***	***Sundatrigona n. gen.***
1975	Sakagami	*Trigona (Lepidotrigona) ventralis*" forma *nigribasis* nov.	nomen nudum
1978	Sakagami	*Trigona (Tetragonula) gressitti*	*Tetragonula gressitti*
1978	Sakagami	*Trigona (Tetragonula) hirashimai*	*Tetragonula hirashimai*
1978	Sakagami	*Trigona (Tetragonula) pagdeniformis*	*Tetragonula pagdeniformis*
1978	Sakagami	*Trigona (Tetragonula) zucchii*	*Tetragonula zucchii*

1978	Sakagami	*Trigona (Tetragonula) latigenalis*	*Tetragonula reepeni*
1978	Sakagami	*Trigona (Tetragonula) minor*	*Tetragonula minor*
1982	Michener	***Proplebeia***	
1985	Sakagami & Inoue	*Trigona (Tetragonula) minangkabau f. darek*	Nomen nudum
1985	Sakagami & Inoue	*Trigona (Tetragonula) minangkabau*	*Tetragonula minangkabau*
1987	Sakagami & Inoue	*Trigona (Trigonella) lieftincki*	*Sundatrigona lieftincki*
1989	Sakagami & Inoue	*Geniotrigona incisa*	*Wallacetrigona incisa*
1990	Michener & Sakagami	***Papuatrigona***	
1993	Inoue & Sakagami	***Sundatrigona***	
1998	Tadauchi et al	***Pacificotrigona*** (Nomen Nudum)	***Sahulotrigona n. subgen.***
1998	Tadauchi et al	*Trigona (Pacificotrigona) okazawai irianensis*	*Trigona (Pacificotrigona) okazawai*
1998	Tadauchi et al	*Trigona wallacei*	*Nomen Nudum*
1998	Tadauchi et al.	*Trigona bismarkiana*	*Nomen nudum*
2000	Engel	*Lisotrigona carpenteri*	*Lisotrigona carpenteri*
2000	Engel	*Lisotrigona furva*	*Lisotrigona furva*
2004	Franck	*Trigona davenporti*	*Tetragonula davenporti*
2004	Jobiraj & Edo	*Lisotrigona mohandasi*	*Lisotrigona mohandasi*
2004	Michener & Boongird	*Trigona (Tetragonula) sirindhornae*	*Tetragonula sirindhornae*
2015	Dollin, Dollin & Rasmussen	*Austroplebeia magna*	*Austroplebeia magna*
2017	Rasmussen et al.	***Sahulotrigona***	n. subgen.
2017	Rasmussen & Engel	***Wallacetrigona***	n. gen.
2017	Viraktamath and Sajan Jose	*Lisotrigona chandrai*	*Lisotrigona chandrai*
2017	Viraktamath and Sajan Jose	*Lisotrigona revanai*	*Lisotrigona revanai*
2017	Rasmussen et al.	*Sahulotrigona paradisaea*	n. sp.
2017	Engel et al	*Tetragonula malaipanae*	n. sp.

13.6 Apiculture to Meliponiculture Historical evolution.

Figure 157 The Beekeepers, 1568, by Pieter Bruegel the Elder

Humans have collected honey for at least 15 thousand years, indicated by cave wall depictions. Bee rearing probably started some 9 thousand years ago as found in dated pottery used in North Africa. Later, bee domestication for honey and pollination in its crude form must have started during the Pharaohs some 4 to 5 thousand years ago, as seen in hieroglyphs and evident in stored jars of honey found in pyramid tombs of the Pharoahs. (Wikipedia, 2016)
Illustrations in a Medieval health handbook showed some form of Apiculture as far back ad the 14th century. It wasn't until the 18th century that European understanding of the colonies and biology of bees allowed the construction of the moveable comb hive to harvest honey without destroying the entire colony.

However, the Mayans have been rearing stingless bees a few thousand years before this. (Roubik 2005). It is from these practices Meliponiculture evolved. In the Maya region, the bee of choice was *Melipona beecheii*, called xuna'an kab or colel-kab ("royal lady") in the Maya language. Bishop Diego de Landa describes the beekeeping practices of domesticated *Melipona beecheii* (the most-used stingless honey bee) and the practices of hunting and collecting honey from wild bees in the forests. (Imre, Young, & Marcus, n.d.)

The Mayans have been at it (Meliponiculture) in the last 2,500 years ago. I refer to a Video titled "Honey for the Maya", a short film by Dr. Stephen Buchmann of the Drylands Institute, Tucson, AZ (https://www.youtube.com/watch?v=d_pjoDxwYS8). It is part of a project he is fundraising for. "I would like to finish this project before Dec. 21, 2012". Looking at this dreaded date, I scoured the other videos and one by Elisabeth Theriot. Aptly titled "2012: Why Did the Mayans Predict Armageddon? | Mayan Revelations: Decoding Baqtun" (https://www.youtube.com/watch?v=gK5YJO8wb7w). In this video, she embarks on an epic journey of discovery to dispel the myths about "the December 21st 2012 end of the world" conspiracy theories surrounding Aztec and Mayan Calendars.

One particular issue that fascinated me was the invention of zero. "The zero has a value for Mayans, unlike other cultures where zero means nothing and is empty. For example, in many Estalas (**monuments fashioned by the Maya civilization** of ancient Mesoamerica), we see a flower. As the Mayans call it, it is the calendared flower, meaning zero.

Figure 158 The placenta in many Mayan languages is "flower (Kozij)".

Figure 159 The Mayan zero is also depicted with a seashell or snail.

The placenta in many Mayan languages is called "flower (Kozij)". It is the principle of origin. The Mayan zero is also depicted with a seashell or snail. That symbolizes creation and the spiral of time. The beginning and the infinite.

The Mayan Abuelos claim the spiral is the form of creation and a manifestation of reality. So, our DNA consists of a spiral shape of space-time; the Naj, a macro spiral, is the ratio of all creation. I am just exploring the spiral of the Meliponine brood cell arrangement in other cultures and its occurrence in nature. The Mayan zero depiction is that of a left-handed mollusc. Maybe just a coincidence the spiral indicates zero in Mayan culture. This correlation of the seashell spiral to the spiral of the Meliponine brood structure is explored in item 9.1.1 (Right-handedness Trivia) page 69 (Figure 106 Sinistral and dextral coils of molluscs.).

13.6.1 Precolumbian Uses of Bees

The domestic use of stingless bees in Mexico dates from pre-Columbian times. This activity reached a particularly impressive level in the Yucatan peninsula, where the Mayan people developed meliponiculture to a level similar to that of management of honey bees during Medieval times in Europe (Quezada-Euán et al., 2001). Accounts on ancient meliponiculture are not abundant in the Americas. Most pre-Columbian evidence comes from Mexico, in particular from the Yucatan Peninsula. Early accounts from Spaniards indicate that outstanding levels of farming and large concentrations of colonies of *Melipona beecheii* ('Xunan kab' in Mayan) were present in the region. An extensive and probably monopolized trade of honey and cerumen with distant territories may have supported the economy and the development of

Mayan civilization (Quezada-Euán, 2018). The products of bees, honey and wax[15], were used in Precolumbian Mesoamerica for religious ceremonies, medicinal purposes, as a sweetener, and to make the hallucinogenic honey mead called balche (Hirst, 2016).

13.6.2 Honey Hunting & Beekeeping

Figure 160 Beekeepers, from "Venationes, Ferrum, Avium, Piscium" (Of Hunting: Wild Beasts, Birds, Fish), engraved by Jan Collaert (1566–1628).

Early humans relied on hunting wild animals and gathering vegetables and fruits, and while their hunter-gatherer lifestyle, they would have come across honey in bees' nests high in the trees. Bees provided our ancestors with perhaps their first condiment—honey. This complex concoction was alluring, and its desire drove men to work in groups to capture the golden prize. Their stories are documented in art, and cave drawings from the late Stone Age show a deeply rooted association between humans and bees. Artworks dating to thirteen thousand years depict amazing feats, with men scaling impossibly tall trees, risking falls and stings, and passing the sweet comb down to helpers below. At first, this bravery was the only way of harvesting honey, and it was several thousand years before the refined practice of beekeeping was developed.

13.6.3 Ancient Egypt

Some three thousand years old, the earliest known human-constructed beehive was discovered in Israel, but the Ancient Egyptians were the first known beekeepers. Evidence from cave drawings in Egypt reveals a long history of beekeeping, dating to at least 2400 BCE and thought to go back as far as 5000 BCE. At first, the Egyptians relied on wild bees' nests, but their beekeeping practices later became very advanced. Not only did they construct beehives in the form of woven baskets covered in clay, which stayed perma- gently

[15] Editor's Note: Not really. The church needed wax candles, which do not drip. Sustainable was not yet invented, but they certainly were not thinking about food, other than honey.

Figure 162 Drawing from ancient Greece of a woman with some honeycomb.

Figure 162 Detail of a bee from a mural in the tomb of Seti I of the 19th Dynasty, from the Valley of the Kings, Luxor, Egypt

In one location, but they also used migratory hives, which floated down the River Nile on rafts, producing unique blends of honey as the bees visited the ever-changing riverside flowers. Egyptians favoured the honey bees that were the best honey producers in the local.

Environment: Through artificial selection, they gave evolution a helping hand to produce a new subspecies, the Egyptian honey bee (*Apis mellifera lamarckii*). Some of the beekeepers were likely a lower class of workers, who were forced to work with these aggressive bees and return glorious honey to their superiors and eventually to the pharaohs and the gods. Official guards accompanied other honey harvesters to explore deep into the sur- rounding land, searching for wild hives, low-lying bushes and tall trees. (Wilson-Rich et al. 2014)

13.6.4 Mesoamerica

The people of Mesoamerica (Mexico, Belize, Guatemala, El Salvador, Costa Rica, Honduras, and Nicaragua) kept stingless bees for some two thousand years before Europeans brought the western honey bee to their shores. There are likely at least 250 stingless bee species in Brazil and at least 4,000 bee species native to North America alone, but the western honey bee is not native to the Americas[16]. The practice of beekeeping with honey bees was introduced to the New World by Europeans in the seventeenth century as a means of sustainable food production. Some honey bee colonies escaped from their beekeepers' management and flew off into the woods in this unexplored new land. Native Americans named these feral honey bees "white man's flies," as their arrival announced the advance of the settlers and gave warning of the in- evitable conflict over land that lay ahead. (Wilson-Rich et al. 2014)

13.6.4.1 Historical review:

The "meliponicultura" in Mesoamerica was a very important activity for the prehispanic settlers since everything they obtained from the beehives was considered mainly for medicinal use.

Figure 163 Mexican Stingless bees by Dr. Margarita Medina Camacho in Mexico City, Mexico (with permission 2014)

[16] Editor's Note: You kind of forget the later arrival of Africanized honeybee form Africa, which arrived in Panama in 1984 and in Central America in 1985, where there were NEVER honey bees of any kind living in the wild (by honey bees here, I mean genus Apis).

The Spanish conquest led to the almost total disappearance of this activity, but some people did not lose their cultural habits and managed to continue breeding stingless bees in clay pots or, as they are commonly called, "Cufflinks".

Reference of the places where they still have this type of stingless bees:
In Mexico, this activity survived over the years on the Yucatan Peninsula, Veracruz and the Sierra Norte de Puebla. The production continues as it did ancestrally. (Pers. Comm. Bernasconi of Universidad de la Vida Diaria, Argentina 2017)

13.6.5 Ancient Bees of the Americas

Advent of Indo-Malayan / Australasian Stingless Bee Taxonomy - Meliponine Taxa by

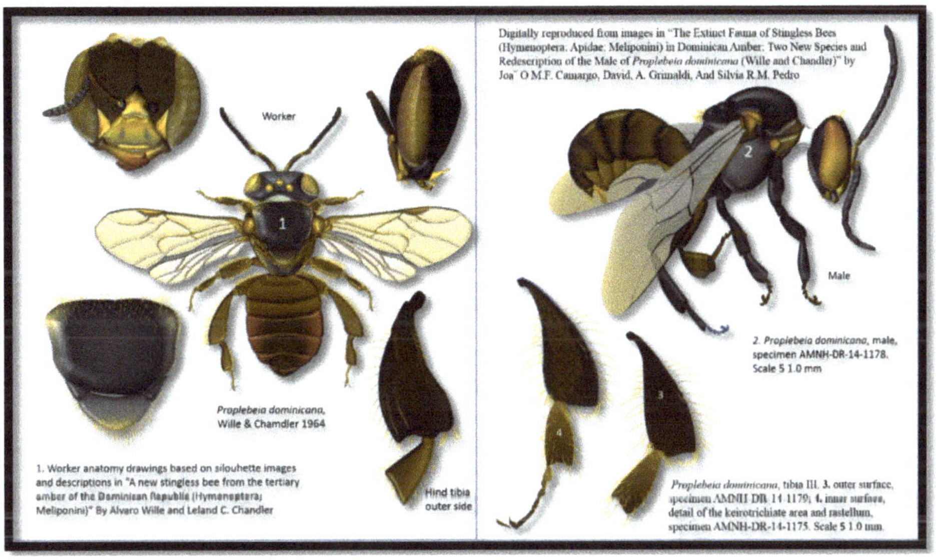

Figure 164 Impressions of Fossilized Bees – Digital reconstruction of the ancient meliponines

13.7 Prominent Authors

Portraits of authors

Table 6 Names Authored by Charles Bingham in 1897 - 1903

Bingham	***Melipona basimaculata***	**1903**	**Nomen dubium (Rasmussen et al. 2017)**
Bingham	*Melipona lutea*	1897	*Homotrigona lutea* (Bingham 1897)
Bingham	*Melipona smithii*	1897	*Trigona ruficornis* Smith
Bingham	*Melipona castanea*	1903	*Homotrigona fimbriata* (Smith 1857)

13.7.1 Charles Thomas Bingham (16/4/1848, Ireland – 18/10/1908 West Kensington, London) was an Irish military officer and entomologist. (Wikipedia, Charles Thomas Bingham , 2013)

He named four stingless bees between 1897 to 1903. His tendencies were colour attributes, e.g., *basimaculata, lutea* & *castanea.* Interestingly he named one patronym dedicated to Smith.

smithii, *Melipona* Bingham 1897 [Unnecessary replacement name for *Trigona ruficornis* Smith][#] - Named after the entomologist F. Smith

Portrait 2 Charles Thomas Bingham Created: January 1, 1908 (Wikipedia, Charles Thomas Bingham, 2013)

13.7.2 Frederick Smith - (Born 30/12/1805 – Died 16/2/1879) was a British entomologist. Smith worked in the zoology department of the British Museum from 1849, specializing in the Hymenoptera. In 1875 he was promoted to Assistant Keeper of Zoology. His publications included *Catalogue of Hymenopterous Insects* (seven parts, 1853–1859) and parts 5 (1851) and 6 (1852) of the *Nomenclature of Coleopterous Insects.* (Wikipedia, Frederick Smith, 2013)

Table 7 Names authored by Frederick Smith in 1857 - 1878

Smith	***Trigona collina***	**1857**	***Tetragonilla collina*** **(Smith 1857)**
Smith	*Trigona apicalis*	1857	*Tetrigona apicalis* (Smith 1857)
Smith	*Trigona atricornis*	1865	*Platytrigona atricornis*
Smith	*Trigona atripes*	1857	*Tetragonilla atripes*
Smith	*Trigona canifrons*	1857	*Lophotrigona canifrons* (Smith 1857)
Smith	*Trigona carbonaria*	1854	*Tetragonula carbonaria* (Smith 1854)
Smith	*Trigona fimbriata*	1857	*Homotrigona fimbriata* (Smith 1857)
Smith	*Trigona iridipennis*	1854	*Tetragonula iridipennis*
Smith	*Trigona laeviceps*	1857	*Tetragonula laeviceps*
Smith	*Trigona nitidiventris*	1857	*Lepidotrigona nitidiventris* (Smith 1857)
Smith	*Trigona planifrons*	1865	*Platytrigona planifrons*
Smith	*Trigona terminata*	1878	*Lepidotrigona terminata* (Smith 1857)
Smith	*Trigona thoracica*	1857	*Geniotrigona thoracica* (Smith 1857)
Smith	*Trigona ventralis*	1857	*Lepidotrigona ventralis*
Smith in Horne & Smith,	*Melipona ruficornis*	1870	*Tetragonula ruficornis*

13.7.2.1 Smith authored 15 names consistently between 1854 and 1878 (one year before his passing). They all were anatomical colour and physical attributes except for one ***collīna,*** *Tetragonilla* (Smith, 1857) [Typ. loc.: Malaysia "Malacca (**Mount** Ophir)"][#] (Rasmussen,

2008) = From Latin - **collīnus** m (*feminine* **collīna**, *neuter* **collīnum**); *first/second declension* - Of a hill; growing on a hill. **More commonly known by its Malay name, Gunung Ledang, Mount Ophir** is a mountain in the Gunung Ledang National Park located in Ledang District (northwestern Johor), Malaysia. The summit is located between the border of Muar and Malacca.

13.7.3 Dr. Jesus Santiago Moure* The first voyages made by Padre Moure abroad occurred in the early 1950s when visiting Argentine institutions. During the stay of Prof. Michener in Brazil, they made a trip to Argentina and Chile. Father Moure toured throughout Brazil for many years, ministering his famous Numerical Taxonomy courses for students in various research centres and during congresses, mainly of Zoology. (Databank, Padre Jesus Santiago Moure, 2013)

13.7.3.1 All those species taxa (up to 1960) remain pertinent to this date, albeit the genus was rearranged in 1961 by Moure.

Table 8 Meliponine Genus names authored by Jesus Santiago Moure in 1961

Moure	***Austroplebeia***	**1961**
Moure	***Geniotrigona***	1961
Moure	***Homotrigona***	1961
Moure	***Lisotrigona***	1961
Moure	***Lophotrigona***	1961
Moure	***Odontotrigona***	1961
Moure	***Pariotrigona***	1961
Moure	***Platytrigona***	1961
Moure	***Tetragonilla***	1961
Moure	***Tetragonula***	1961
Moure	***Tetrigona***	1961

Portrait 3 Prof Jesus Santiago Moure (Moure's Bee Catalogue, 2008)

13.7.3.2 Other genera Patronyms by Moure as follows:

13.7.3.2.1 ***Camargoia*** Moure, 1989 = [homage to Prof. João Maria Franco de Camargo.]* (Moure's Bee Catalogue, 2008)

João Maria Franco Camargo: A Brazilian ecologist and bee taxonomist working on Tribe Meliponini. Portrait 4 (Wikia, João M. F. de Camargo, 2007)

13.7.3.2.2 ***Duckeola*** Moure, ***Duckeola*** Moure, 1944 = patronym for Adolf **Ducke** (Moure's Bee Catalogue, 2008) + ola (see suffix **-ola**)

Adolpho Ducke (19/10/1876 in Trieste, Italy – 5/1/1959 in Fortaleza, Brazil), also referred to as **Adolfo Ducke** and occasionally misspelt **"Duque"**, was a notable entomologist, botanist and ethnographer of Amazonia. (Wikipedia, Adolpho Ducke, 2014)

Portrait 4 Prof. João Maria Franco de Camargo ((Wikia, João M. F. de Camargo, 2007))

13.7.3.2.3 ***Friesella*** Moure, 1946 = honouring H. Friese (Moure's Bee Catalogue, 2008) + lla (see suffix –lla).; see also *Frieseomelitta*

Heinrich Friedrich August Karl Ludwig Friese (1860–1948) was an important pioneer bee biologist (melittologist). Between 1883 and 1939, he described 1,989 new species and 564 new varieties or subspecies of insects, of which over 99% were bees. Portrait 5 (Claus Rasmussen & John S. Ascher, 2008)

Portrait 5 Heinrich Friese. (Claus Rasmussen & John S. Ascher, 2008)

13.7.3.2.4 ***Kelnermelia*** Moure *In* Moure and Camargo, 1978: The generic name is in homage to Dr. S. Kelner-Pillault. + melia (Greek μέλια "honey [&&] (Moure J.S. & Camargo J.M.F., 1978)

13.7.3.2.5 ***Melikerria*** Moure, (1992). = Latin Meli (honey) + kerria [homage to Dr. Warwik E. Kerr.]* (Moure's Bee Catalogue, 2008)

Warwick Estevam Kerr The agronomist, geneticist and biologist Warwick Kerr are considered the foremost expert in the genetics of bees in the world. Portrait 6 (Databank, Warwick Estevam Kerr., 2013)

Portrait 6 Warwick Kerr 1922 (Wikia, Warwick Kerr)

13.7.3.2.6 ***Michmelia*** Moure, (1975). = mich melia (Latin phrase for "much better"); Relates to the subgenus as in *Melipona (Michmelia)* Moure, 1975 - honouring Prof. C. D. Michener (See Portrait 7) + melia (Greek μέλια "honey")

13.7.3.2.7 ***Micheneria*** Kerr, et al. (1967). = [homage to Prof. C. D. Michener.]* (Moure's Bee Catalogue, 2008) + ia (see suffix –ia)

The American entomologist Charles Michener was born in Pasadena, California, on 22/9/1918. Much of his career has been devoted to bees' systematics and natural history. Portrait 7 (Wikipedia, Charles Duncan Michener, 2014)

13.7.3.2.8 ***Mourella*** Schwarz in Moure, (1946) pp.442-443 (see Dr. Jesus Santiago Moure, Portrait 3 **Prof Jesus Santiago Moure (Moure's Bee Catalogue, 2008)** Introduced first as a subgenus of *Trigona* Jurine 1807

Portrait 7 Charles D. Michener Photo: Deborah Smith

13.7.3.2.9 ***Nogueirapis*** Moure, 1953 = Named after the entomologist Paulo Nogueira-Neto (Moure's Bee Catalogue, 2008)+ apis (Latin – "bee")

Paulo Nogueira-Neto Professor of Ecology, Institute of Biosciences AP, University of São Paulo, Brazil Honorary Professor of the Institute of Studies Forward, University of São Paulo Portrait 8 (Nogueira-Neto, 1997)

Portrait 8 Paolo Nogueira-Neto (Nogueira-Neto, 1997)

13.7.3.2.10 ***Sakagamilla*** Moure, (1989) = [honouring Soichi F. Sakagami]* (Moure's Bee Catalogue, 2008) + -illa (see suffix **–illa**)

Shôichi Francisco **Sakagami** (Born 4/1/1927 at Tokyo's Shiba – Died 4/11/1996 in Sapporo) Specialty in Social insects Portrait 9 (Senckenberg Biographies, 2013)

13.7.3.2.11 ***Schwarziana*** Moure, 1943 = Named after the entomologist Herbert F. Schwarz (Moure's Bee Catalogue, 2008) + iana (see suffix –iana) (see also *Schwarzula*)

Portrait 9 Shoichi Francisco Sakagami (Kevan P & Imperatriz Fonseca VL (eds), 2002)

13.7.3.2.12 ***Schwarzula*** Moure, 1946 = Named after the entomologist Herbert F. Schwarz (Moure's Bee Catalogue, 2008) + ula (see suffix ula) (see also *Schwarziana*)

Herbert Ferlando Schwarz, American entomologist, specializing in Hymenoptera, mostly in stingless bees, Meliponini Tribe. He described numerous genera and species. Schwarz was working in the section on insects and spiders the American Museum of Natural History (American Museum of Natural History) (Wikia, Herbert Ferlando Schwarz, 2014)

13.7.3.3 Other Meliponini genera authored by Dr. Jesus Santiago Moure as follows:

Aparatrigona Moure, 1951, ***Apotrigona*** Moure, 1961, ***Celetrigona*** Moure, 1950, ***Cleptotrigona*** Moure 1961, ***Dolichotrigona*** Moure, ***Eomelipona*** Moure, (1992), ***Geotrigona*** Moure, 1943 (Introduced first as a subgenus of *Trigona* Jurine 1807.), ***Leurotrigona*** Moure, 1950 (first introduced as a subgenus of *Hypotrigona* Cockerell, 1934), ***Liotrigona*** Moure, ***Plebeina*** Moure, 1961, ***Plebeiella*** Moure, 1961, ***Ptilotrigona*** Moure, 1951, ***Scaptotrigona*** Moure, (1942) (Introduced first as a subgenus of *Trigona*. Jurine, 1807) Tetragonisca Moure, 1946, ***Trichotrigona*** Camargo & Moure, 1983, ***Trigonisca*** Moure, 1950

13.7.4 Charles Duncan Michener, besides authoring "The Bees of The World" in 2000, also named the following in the Indo-Malayan/Australasian Clade:

Table 9 Meliponini Names authored by Michener in 1990 - 2004

Michener & Sakagami	***Papuatrigona***	**1990**
Michener	***Proplebeia***	1982
Michener & Boongird	*Trigona (Tetragonula) sirindhornae*	2004

Before this, he named three species, *Liotrigona* ***madecassa*** (Brooks & Michener 1988), *L.* ***bitika*** (Brooks & Michener 1988) and *L.* ***mahafalya*** (Brooks & Michener 1988), from Madagascar. Then *Liotrigona* ***nilssoni***, Michener, 1990 and more recently he co-authored an extinct genus and its' species ***Exebotrigona velteni***, (Engel & Michener 2013). *Trigona* ***pseudojatiformis*** Wille. & Michener (1973) and **b**efore this, there was *Hypotrigona* ***araujoi*** (Michener 1959)

13.7.4.1 ***Proplebeia*** Michener, (1982). = [An extinct genus *Proplebeia*, recognizes four species, three of them from the amber of Dominican Republic, and one from the amber of Chiapas, Mexico, dating from the Miocene (see Camargo *et al*., 2000)]* (Moure's Bee Catalogue, 2008) Prefix Pro- ("earlier") + plebeia (see **plebeia**) (first introduced as a subgenus of *Trigona* Jurine, 1807).

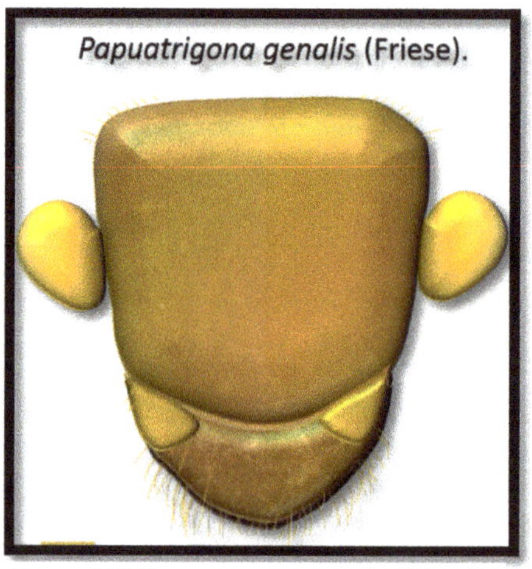

Figure 165 Papuatrigona genalis (Friese)

13.7.4.2 ***sirindhornae,*** *Tetragonula* (Michener & Boongird, 2004**)** [Etym.: This species was dedicated to HRH Princess Sirindhorn of Thailand on the occasion of her 48th birthday celebrations]`` (Boongird, 2004)

13.7.4.4 ***Papuatrigona*** Michener & Sakagami, 1990 (Rasmussen, 2008) = Pertaining to Papua New Guinea. + trigona (see suffix –trigona)

Papua (**Indonesian**: *West Papua*) is a province of Indonesia. It comprises most of the western half of the island of **New Guinea** and nearby islands. (Wikipedia, West Papua, 2014)

13.7.5 Giovanni Gribodo, the entomologist, brings us back to before the 20thcentury**.** Not having much historical data on Gribodo, we find that he describes three names of stingless bees in the region just before the turn of the century in the 1890s.

Table 10 Meliponinin species names authored by Grbodo in 1891 - 1893

Gribodo	*Trigona javanica*	1891	*Lepidotrigona javanica*
Gribodo	*Trigona melanocephala*	1893	*Tetragonula melanocephala* (Gribodo 1893)
Gribodo	*Trigona melina*	1893	*Tetragonula melina*

13.7.6 Heinrich Friedrich August Karl Ludwig Friese, (Portrait 5) however, at the turn of the century, a prolific author for bees with more than 2,000 names of species and varieties or subspecies to his credit. Sixteen names of Indo-Australasian stingless bee names are well diversified in the description. We have patronyms (e.g., *reepeni, biroi, pfeifferi*), names of locality (e.g., *australis, keyensis, borneensis*), anatomical attributes (e.g., *pygmaea, genalis, latipes*), colour attributes (e.g., *flaviventris, luteiventris, versicolor* & etc.).

Table 11 Meliponine species names authored by Friese in 1898 - 1933

Friese	*Trigona biroi*	1898	*Tetragonula biroi*
Friese	*Trigona mellipes*	1898	*Tetragonula mellipes*
Friese	*Trigona australis*	1898	*Austroplebeia australis*
Friese	*Trigona latipes*	1900	*Lepidotrigona latipēs*
Friese	*Trigona keyensis*	1901	*Platytrigona keyensis*
Friese	*Trigona flaviventris*	1909	*Platytrigona flaviventris*
Friese	*Trigona genalis*	1909	*Papuatrigona genalis*
Friese	*Trigona laeviceps var. clypearis*	1909	*Tetragonula clypearis (Friese 1909)*
Friese	*Trigona versicolor*	1909	*Homotrigona fimbriata (Smith 1857)*
Friese	*Trigona anamitica*	1909	*Homotrigona anamitica (Friese 1909)*
Friese	*Trigona reepeni*	1918	*Tetragonula reepeni (Friese 1918)*
Friese	*Trigona pfeifferi*	1925	*Tetragonula fuscobalteata (Cameron 1908)*
Friese	*Trigona borneënsis*	1933	*Geniotrigona lacteifasciata*
Friese	*Trigona pygmaea*	1933	*Tetragonula fuscobalteata (Cameron 1908)*
Friese	*Trigona sericea*	1933	*Tetrigona apicalis (Smith 1857)*

13.7.7 Percy Tarlton Rayment FRZS (November 27 1882 – June 17 1964) was an Australian artist, author, broadcaster, poet, naturalist, entomologist and beekeeper. He is especially renowned for his extensive pioneering studies of Australia's native bees. (http://adb.anu.edu.au/biography/rayment-percy-tarlton-8164)

Table 12 Meliponinin species names authored by Rayment in 1930 - 1932

Rayment	*Trigona cockerelli ornata* subsp.	1932	*Trigona (Heterotrigona) davenporti (Franck, 2004)*
Rayment	*Trigona symei*	1932	*Austroplebeia symei*
Rayment	*Trigona websteri*	1932	*Austroplebeia websteri*
Rayment	*Trigona cockerelli ornata*	1932	*Austroplebeia ornata* (Rayment 1932)
Rayment	*Trigona cockerelli*	1930	*Austroplebeia cockerelli*

13.7.8 Theodore Dru Alison Cockerell (1866–1948) was an American zoologist, born at Norwood, England, and brother of Sydney Cockerell. He was educated at the Middlesex and then studied botany in Colorado in 1887-90. Subsequently, he became a taxonomist and published numerous papers on the Hymenoptera, Hemiptera, Mollusca, and palaeontology and evolution publications. Portrait 10 (Wikipedia, Theodore Dru Alison Cockerell , 2013)

Portrait 10 T.D.A. Cockerell (Wikipedia, Theodore Dru Alison Cockerell, 2013)

Table 13 Taxa authored by T.D.A Cockerell: 1918 -1934

Cockerell	*Hypotrigona*	1934	
Cockerell	*Trigona aliceae*	1929	*Homotrigona aliceae* (Cockerell 1929)
Cockerell	*Trigona bakeri*	1919	*Heterotrigona bakeri*
Cockerell	*Trigona carbonaria hockingsi* n. subsp.	1929	*Tetragonula hockingsi*
Cockerell	*Trigona valdezi*	1918	*Tetragonula valdezi*
Cockerell	*Trigona dapitanensis*	1925	*Tetragonula dapitanensis*
Cockerell	*Trigona essingtoni*	1905	*Austroplebeia essingtoni*
Cockerell	*Trigona lamingtonia*	1929	*Platytrigona lamingtonia*
Cockerell	*Trigona palavanica*	1915	*Lepidotrigona palavanica*
Cockerell	*Trigona penangensis*	1919	*Tetragonula penangensis*
Cockerell	*Trigona apicalis peninsularis* subsp.n.	1927	*Tetrigona peninsularis*
Cockerell	*Trigona sandacana*	1919	*Heterotrigona erythrogastra* (Cameron 1902)
Cockerell	*Trigona cassiae*	1910	*Austroplebeia cassiae*
Cockerell	*Trigona flavibasis*	1929	*Lepidotrigona flavibasis*
Cockerell	*Trigona fuscibasis*	1920	*Tetragonilla fuscibasis*
Cockerell	*Trigona hæmatoptera*	1919	*Odontotrigona haematoptera*
Cockerell	*Trigona itama*	1918	*Heterotrigona itama* (Cockerell 1918)

Cockerell	*Trigona melanoleuca*	1929	*Tetrigona melanoleuca*
Cockerell	*Trigona cincta percincta*	1929	*Austroplebeia percincta*
Cockerell	*Trigona rufibasalis*	1918	*Tetragonilla rufibasalis*
Cockerell	*Trigona trochanterica*	1920	*Lepidotrigona trochanterica*
Cockerell	*Trigona arcifera*	1929	*Lepidotrigona arcifera* (Cockerell 1929)
Cockerell	*Trigona (Tetragonula) devicta*;	1921	*Tetragonula devicta*
Cockerell	*Trigona sapiens*	1911	*Tetragonula sapiens*
Cockerell	*Trigona geissleri*	1918	*Tetragonula geissleri* (Cockerell 1918)
Cockerell	*Trigona cambodiensis*	1926	*Tetragonilla collina* (Smith 1857)
Cockerell	*Trigona wybenica*	1929	*Tetragonula clypearis* (Friese 1909)
Cockerell	*Trigona angophorae*	1912	*Tetragonula carbonaria* (Smith 1854)
Cockerell	*Trigona ambusta*	1918	*Geniotrigona thoracica* (Smith 1857)
Cockerell	*Trigona breviceps*	1919	*Heterotrigona itama* (Cockerell 1918)
Cockerell	*Trigona ferrea*	1929	*Homotrigona lutea* (Bingham 1897)
Cockerell	*Trigona fulvomarginata*	1919	*Lepidotrigona terminata* (Smith 1857)
Cockerell	*Trigona hemileuca*	1929	*Tetrigona apicalis* (Smith 1857)
Cockerell	*Trigona melanotricha*	1918	*Homotrigona anamitica* (Friese 1909)
Cockerell	*Trigona pallidicincta*	1918	Incerta sedis
Cockerell	*Trigona scintillans*	1920	*Lisotrigona cacciae* (Nurse, 1907)
Cockerell	*Trigona atomella*	1919	*Tetragonula fuscobalteata* (Cameron 1908)
Cockerell	*Trigona busara*	1918	*Lophotrigona canifrons* (Smith 1857)
Cockerell	*Trigona confusella*	1919	*Tetragonula geissleri* (Cockerell 1918)
Cockerell	*Meliponorytes devictus*	1921	*Trigona (Tetragonula) devicta*;

Interestingly it appears that when naming new taxa of stingless bees, the most often used inspiration is derived from the morphology or the colouration of the bees. This trend is followed in popularity by names proposed in honour of other people and locality names. (Pers. Comm. Rasmussen 2014)

13.7.9 Shôichi Francisco Sakagami did extensive studies on *Tetragonula* in Asia and had predominantly named patronyms and authored and co-authored the following:

Table 14 Meliponini Taxa authored by Sakagami in 1859 - 1989

Sakagami	***Trigona (Tetragonula) hirashimai***	1978	***Tetragonula hirashimai***
Sakagami	*Trigona (Tetragonula) pagdeniformis*	1978	*Tetragonula pagdeniformis*
Sakagami	*Trigona (Tetragonula) zucchii*	1978	*Tetragonula zucchii*
Sakagami	*Trigona (Tetragonula) minor*	1978	*Tetragonula minor*
Sakagami	*Trigona (Tetragonula) gressitti*	1978	*Tetragonula gressitti*
Sakagami	*Trigona (Tetragona) matsumurai*	1959	*Sundatrigona moorei* (Schwarz 1937)

Sakagami	*Trigona (Tetragonula) latigenalis*	1978	*Tetragonula reepeni* (Friese 1918)
Sakagami	*Trigona (Lepidotrigona) ventralis*" forma *nigribasis* nov.	1975	nomen nudum
Sakagami & Inoue	*Trigona (Trigonella) lieftincki*	1987	*Sundatrigona lieftincki*
Sakagami & Inoue	*Trigona (Tetragonula) minangkabau*	1985	*Tetragonula minangkabau*
Sakagami & Inoue	*Geniotrigona incisa*	1989	*Wallacetrigona incisa*
Sakagami & Inoue	*Trigona (Tetragonula) minangkabau* f. *darek*	1985	Nomen nudum
Sakagami et Moure	***Trigonella***	1975	***Sundatrigona*** (1993)

Trigonella was later changed to ***Sundatrigona*** Inoue & Sakagami, 1993 (Rasmussen, 2008) = of the Sunda Islands, a group of islands in the western part of the Malay Archipelago.

13.7.10 Herbert Ferlando Schwarz, from 1937 to 1939, this American entomologist, senior staff in AMNH (See 6.3.2.12) Seems he was concentrating on Indo-Malayan Stingless bee in those three years though he authored many more insects' taxa and books over his lifetime[17].

Table 15 Meliponini names authored by Schwarz in 1937 - 1939

Schwarz	*Plebeia*	1938	
Schwarz	***Heterotrigona***	1939	
Schwarz	***Lepidotrigona***	1939	
Schwarz	*Trigona moorei*	1937	*Sundatrigona moorei* (Schwarz 1937)
Schwarz	*Trigona apicalis var. binghami*	1937	*Tetrigona binghami*
Schwarz	*Trigona haemanoptera var. dulitae*	1937	*Odontotrigona haematoptera*
Schwarz	*Trigona sarawakensis*	1937	*Tetragonula sarawakensis*
Schwarz	*Trigona hobbyi*	1937	*Platytrigona hobbyi*
Schwarz	*Trigona (Tetragona) sarawakensis var drescheri*	1939	*Tetragonula drescheri*
Schwarz	*Trigona (Tetragona) fuscobalteata var. pagdeni*	1939	*Tetragonula pagdeni*
Schwarz	*Trigona (Hypotrigona) pendleburyi*	1939	*Pariotrigona pendleburyi*
Schwarz	*Trigona (Lepidotrigona) ventralis* var. *doipaensis*	1939	*Lepidotrigona doipaensis*
Schwarz	*Trigona (Hypotrigona) pendleburyi var. klossi*	1939	*Pariotrigona klossi*

Herein, the trend was towards the name of places rather than indicating the type locality most of the time, e.g., *sarawakensis, dulitae & doipaensis.*

[17] Editor's note: He overlapped with Michener for a while at AMNH, where Mich's first job was. Schwarz was an MD from a wealthy family (FAO Schwarz toy store in NYC) and he had a hobby of stingless bees. He paid for his own office at AMNH and, on the day he died, the office door blew shut and the glass window in it shattered...

13.7.11 Osamu Tadauchi currently works at the Graduate School of Sciences, Kyushu University. Osamu researches Parasitology, Zoology and Ecology. (https://www.researchgate.net/profile/Osamu_Tadauchi)

Up to as recent as 1998, Tadauchi established names in the Papua region, but the proposed names were invalid because the group designated is not described or illustrated sufficiently for recognition and has no nomenclatural status were considered Nomen Nudum. *Pacificotrigona* was later named by Rasmussen et al as *Sahulotrigona* 2017.

Table 16 Meliponini Names suggested by Tadauchi in 1998

Tadauchi et al	***Pacificotrigona* (Nomen Nudum)**	**1998**	***Sahulotrigona***
Tadauchi et al	*Trigona (Pacificotrigona) okazawai irianensis*	1998	*Trigona (Pacificotrigona) okazawai*
Tadauchi et al	*Trigona wallacei*	1998	*Nomen Nudum*
Tadauchi et al.	*Trigona bismarkiana*	1998	*Nomen nudum*

13.7.12 C. H. Wallace Pugh (1889-1973) Entomologist Specialisms: Diptera**. Pugh** seems to have hurried this batch of names in 1947. One can only speculate what happened just after World War 2.

Table 17 Taxa by Pugh in 1947

Pugh	***Melipona khasiana***	**1947**	***Nomen Nudum***
Pugh	*Melipona tunneli* (Nomen dubium)	1947	*Lepidotrigona arcifera*
Pugh	*Melipona terrestris*	1947	*Nomen dubium*

13.8 Names of Indo-Malayan & Australasian Species & Genus in New Millennium

Table 18 Names of Indo-Malayan & Australasian Species & genus in the New Millennium

Year	Author	Taxa	New Combination
2000	Engel	*Lisotrigona carpenteri*	*Lisotrigona carpenteri*
2000	Engel	*Lisotrigona furva*	*Lisotrigona furva*
2004	Franck	*Trigona davenporti*	*Tetragonula davenporti*
****2004**	Jobiraj & Edo	*Lisotrigona mohandasi*	*syn. Lisotrigona cacciae*
2004	Michener & Boongird	*Trigona (Tetragonula) sirindhornae*	*Tetragonula sirindhornae*
2015	Dollin, Dollin & Rasmussen	*Austroplebeia magna*	*Austroplebeia magna*
2017	Rasmussen et al.	***Sahulotrigona***	n. subgen.
2017	Engel, Michener & Boontop	*Tetragonula (Tetragonula) malaipanae*	n. sp.
2017	Rasmussen & Engel	***Wallacetrigona***	n. gen.

****2017**	Viraktamath and Sajan Jose	*Lisotrigona chandrai*	*syn. Lisotrigona cacciae*
****2017**	Viraktamath and Sajan Jose	*Lisotrigona revanai*	*syn. Lisotrigona cacciae*
2017	Rasmussen et al.	*Sahulotrigona paradisaea*	n. sp.
2017	Engel et al	*Tetragonula malaipanae*	n. sp.

**Note: (*mohandasi, chandrai* and *revanai* in *Lisotrigona* are invalid names (synonyms) and therefore not valid species. The original authors mistakenly described teneral adults of *Lisotrigona cacciae* as separate species, which is a major biological mistake. (Pers. Comm. Engel 2017)[18]

13.8.1 ***Meliponini*** = Translingual – **Etymology:** - *Melipona* (see Melipona) + *-ini* (see suffix –ini) - ***Meliponini*** - A taxonomic **tribe** within the **subfamily Apinae.**

***Remark*:** *Trigonini* is no longer in use (in his "Bees of the World", Michener used only *Meliponini*. Michener, C.D. (2007) [Claus Rasmussen in a personal communication.]

13.8.2 ***furva,*** *Lisotrigona* Engel, 2000 (M. S. Engel 2000) = Latin- **furvus** m (*feminine* **furva**, *neuter* **furvum**); *first/second declension* - dark, dusky, gloomy, swarthy, black.

Etymology: The specific epithet is derived from furvus (Latin, dark) and references the dark colouration without yellow markings. ***Diagnosis***: This species is most similar to *L. scintillans* but differs by the dark integument, strong close mesoscutal punctures, and presence of pubescence on the hypoepimeron.

13.8.3 ***carpenteri,*** *Lisotrigona* Engel 2000 - Relates to collector of *Lisotrigona* ***carpenteri*** Engel, 2000 [Typ. Loc.: Pozo Vietnam: Nghe An Province[19]: Khe Bo, by James Michael **Carpenter** from the AMNH][#] (Rasmussen, 2008)

Etymology: The specific epithet is a patronym honouring Dr. James M. Carpenter (AMNH), who collected the type species and made it available for study. ***Distribution***: This species is known from only two sites in Northern Vietnam at low elevations near the base of the mountain range.

13.8.4 ***davenporti,*** *Tetragonula* (Franck, 2004) also relating to *Trigona (Heterotrigona)* ***davenporti*** **Franck** ***et al.*** 2004 [Typ. Loc.: Australia, Qld, Mudgeeraba, Austinville Road, V. 1990, P. **Davenport** leg."][#] (Rasmussen, 2008)

Remark: This species was dedicated to Peter **Davenport,** who collected the four colonies examined.[1] (Rasmussen, Personal communication, 2013) "Peter Davenport is one of the leading pioneer beekeepers of the Australian stingless bee industry, with a huge collection of stingless beehives and a wealth of experience." (Anne Dollin, Pers. Comm. 2015)

[18] Editor's note: However, now (2021) there are new names invented by Engel (2021) that include these and many other Supraspecific groups

[19] DR. Pham's Note: Change from Nghê An originally, to Nghe An province.

13.8.5 ***mohandasi,*** *Lisotrigona* Jobiraj & Narendran 2004: [Typ. Loc.: India, Kerala Forest Research Institute (Peechi), collected by Dr. K. **Mohandas,**][#] (Rasmussen, 2008) [syn. to *Lisotrigona cacciae* (Nurse 1907)] (Rasmussen 2017)

Discussion: This species resembles *L. cacciae* (Nurse) in general characteristics but differs mainly in the following characteristics: 1. Integument of head and thorax coal-black and metasoma brown; 2. Mandibles dull yellow to brown 3. Gena is sparsely punctate; 4. Hairs on clypeus plumose; 5. Hypoepimeron with pubescence; 6. Metanotum reticulate anteriorly; whereas in *L. cacciae* with the following differences: (1. integument dark brown; 2. mandibles yellow; 3. gena impunctate; 4. Hairs on clypeus simple; 5. Hypoepimeron glabrous; 6. Metanotum finely imbricate) respectively.

Portrait 11
https://en.wikipedia.org/wiki/Alfred_Russel_Wallace

13.8.6 ***sirindhornae,*** *Tetragonula* (Michener & Boongird, 2004) [Etym.: This species was dedicated to HRH Princess Sirindhorn of Thailand on the occasion of her 48th birthday celebrations] (Boongird, 2004)

Note: Princess Maha Chakri Sirindhorn, born Princess Sirindhorn Debaratanasuda Kitivadhanadulsobhak on April 2 1955, is the second daughter of King Bhumibol Adulyadej.

13.8.7 ***magna,*** *Austroplebeia magna* Dollin, Dollin & Rasmussen 2015 [Etymology. The Latin feminine adjective, magna, meaning 'large', refers to the broad basitarsus III and long sting lancet in workers of this species.]

Remarks. The NT populations identified as *A. magna* sp. nov. were part of the 'symei' group defined in our preliminary studies (Halcroft et al. 2015). That group was characterized by its dark colouration. It also included the eastern species, *A. cassiae*. Although superficially similar to *A. cassiae* in size and colouration, A. magna workers could be distinguished by their broad basitarsus III, long sting lancet and fine clypeus hair.

Reference:

Dollin, A. E.; Dollin, L. J.; Rasmussen, C. (2015-11-23). Australian and New Guinean Stingless Bees of the Genus Austroplebeia Moure (Hymenoptera: Apidae)—a revision. Zootaxa. 4047 (1): 1. doi:10.11646/zootaxa.4047.1.1. ISSN 1175-5334.

13.8.9 ***Wallacetrigona*** Rasmussen et al. 2017 [Etymology: The generic name honours Alfred Russel Wallace (1823–1913), Portrait 11, an intrepid and early explorer of the Indomalayan insect fauna and co-author with Charles R. Darwin (1809–1882) of the theory of evolution. The gender of the name is feminine.

Included species: Presently, the genus is understood to only include the type species, *Wallacetrigona incisa* (Sakagami and Inoue), a new combination from Indonesia (Sulawesi), east of the important faunal boundary known as the Wallace Line (and part of the biogeographic area known as Wallacea).

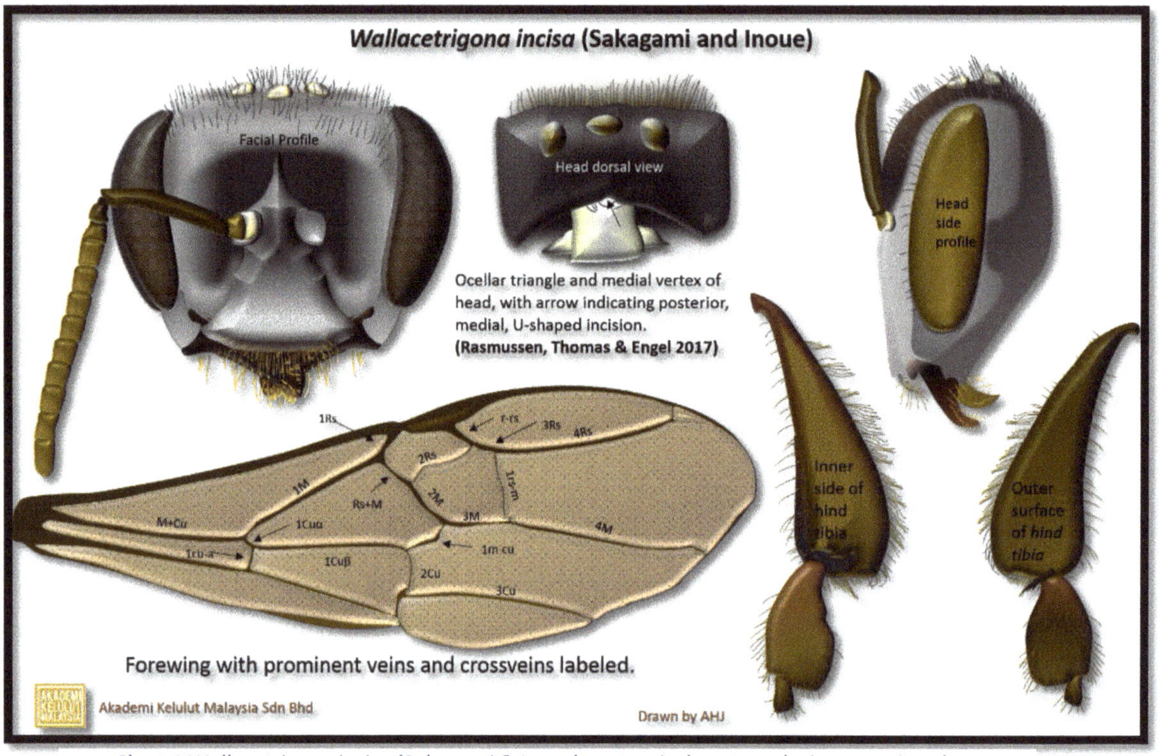

Chart 1 Wallacetrigona incisa (Sakagami & Inoue) anatomical parts and wing venation chart

13.8.10 *Sahulotrigona* Engel & Rasmussen

2017 [Etymology: The subgeneric name is a combination of Sahul, a reference to the occurrence of the known species within a distinct zoogeographic region relative to those *Heterotrigona* occurring in Sundaland, and the generic name *Trigona* Jurine.]

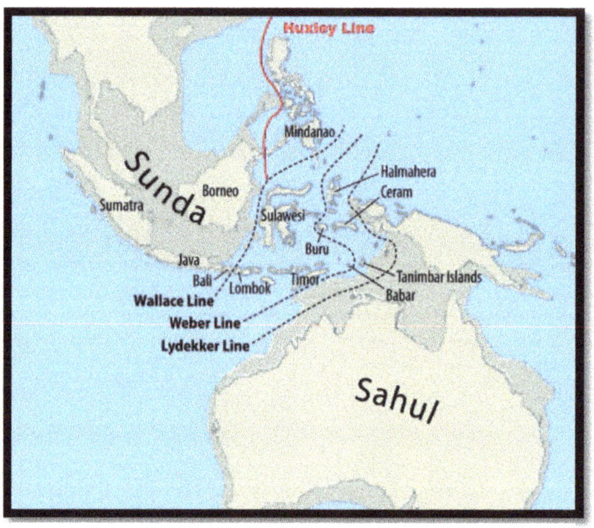

Map 1 Map of Sunda and Sahul - underwater continents

Remark: The gender of the name is feminine. Included species: The subgenus as currently constituted contains only two species, both from New Guinea: *Heterotrigona (Sahulotrigona) paradisaea*, n. sp., and *H. (S.) atricornis* (Smith). (Excerpts from A new subgenus of *Heterotrigona* from New Guinea (Hymenoptera: Apidae), Engel & Rasmussen 2017) Journal of Melittology No. 73 pg. 8. Ref: Source=Base map derived from File: Karte von Sunda und Sahul.png. Map of Sunda and Sahul and the Wallace Line, the Lydekker Line and the Weber Line. Dated, October 1 2012, 13:11:25 by Altaileopard

13.8.11 ***paradisaea**, Heterotrigona (Sahulotrigona) paradisaea* [Etym: The specific epithet is based on the generic name *Paradisaea* Linnaeus (Aves: Passeriformes: Paradisaeidae), the genus of the Emperor bird-of-paradise (*Paradisaea guilielmi* Cabanis), which is the national emblem of Papua New Guinea.

Note: The name is meant to honour the unique biota cultivated by the indigenous peoples of New Guinea.

Figure 166 Emblem of Papua New Guinea

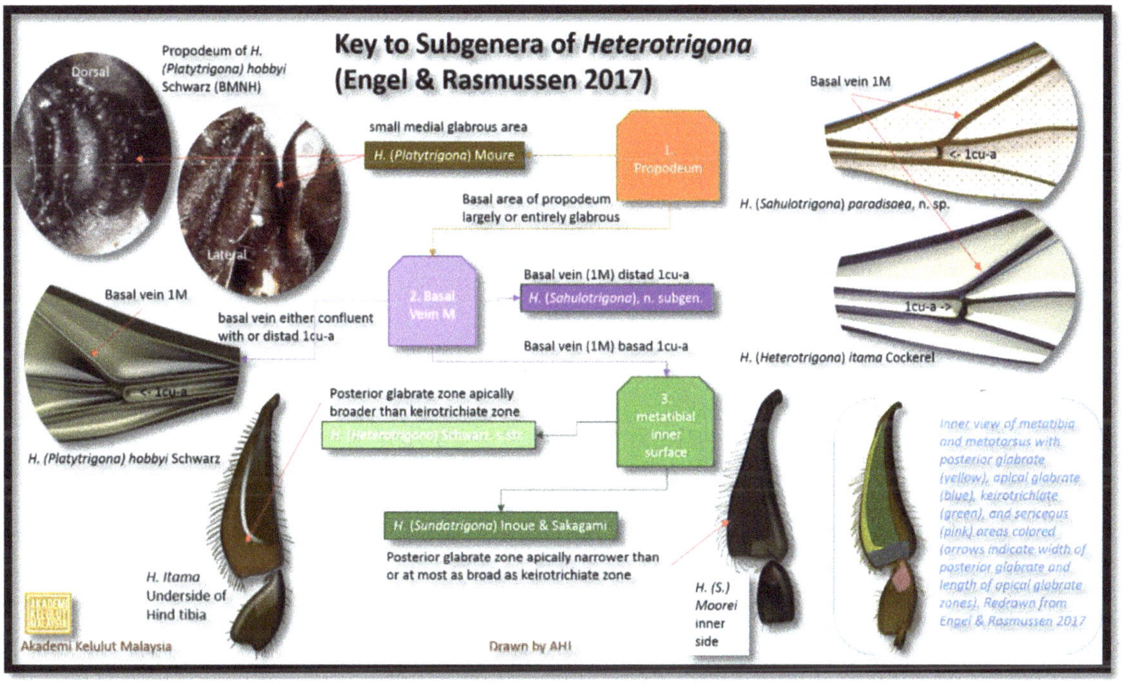

Chart 3 Composite Key of Heterotrigona (Engel & Rasmussen 2017)

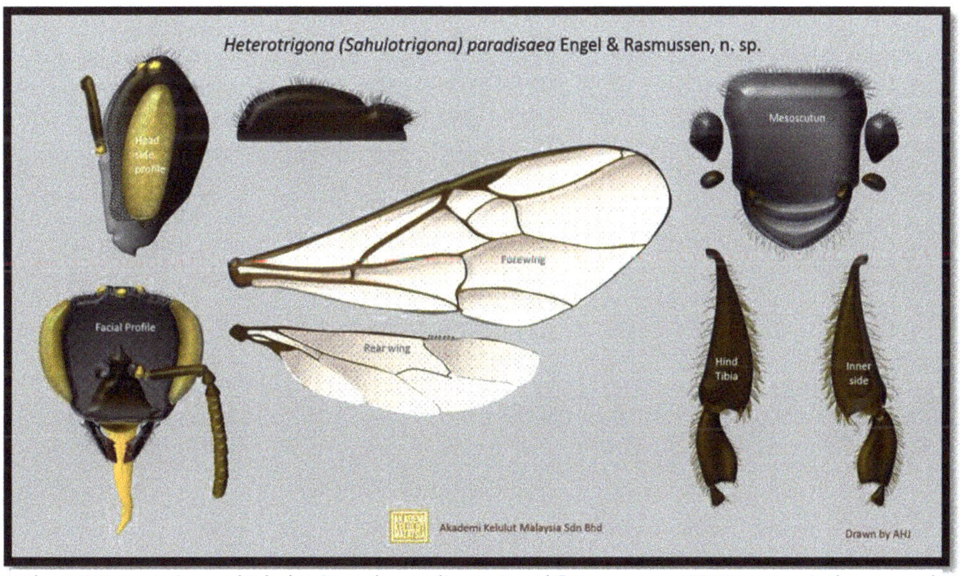

Chart 2 Heterotrigona (Sahulotrigona) paradisaea Engel & Rasmussen 2017 anatomical parts and wing venation chart

13.9 Indo-Malaya – South East Asia

13.9.1 Borneotrigona Engel, new subgenus, Type species: *Trigona hobbyi* Schwarz, 1937.

Etym: The new subgeneric name is a combination of Borneo, the island from which the group is known, and Trigona Jurine. The gender of the name is feminine.

Included species: The subgenus presently includes only the type species, *H. (Borneotrigona) hobbyi* from northern Borneo [Brunei and Malaysia (Sarawak and Sabah)] (Engel 2019).

Reference:

Michael S. Engel, Notes on Papuasian and Malesian stingless bees, with the descriptions of new taxa (Hymenoptera: Apidae) Journal of Melittology No. 88, pp. 1–25. August 12 2019.

Chart 4 Borneotrigona) hobbyi

***13.9.2 malaipanae**, Tetragonula (Tetragonula) malaipanae* Engel, Michener & Boontop 2017 [Etym: The specific epithet honours Savitree Malaipan of the Department of Agriculture, Kasetsart University, Bangkok, beloved mentor of Y.B. and a prominent Thai melittologist.] Holotype: Thailand, Kanchanaburi [Province], Thong Pha Phum [District], 17.ix.2007 [17 September 2007],

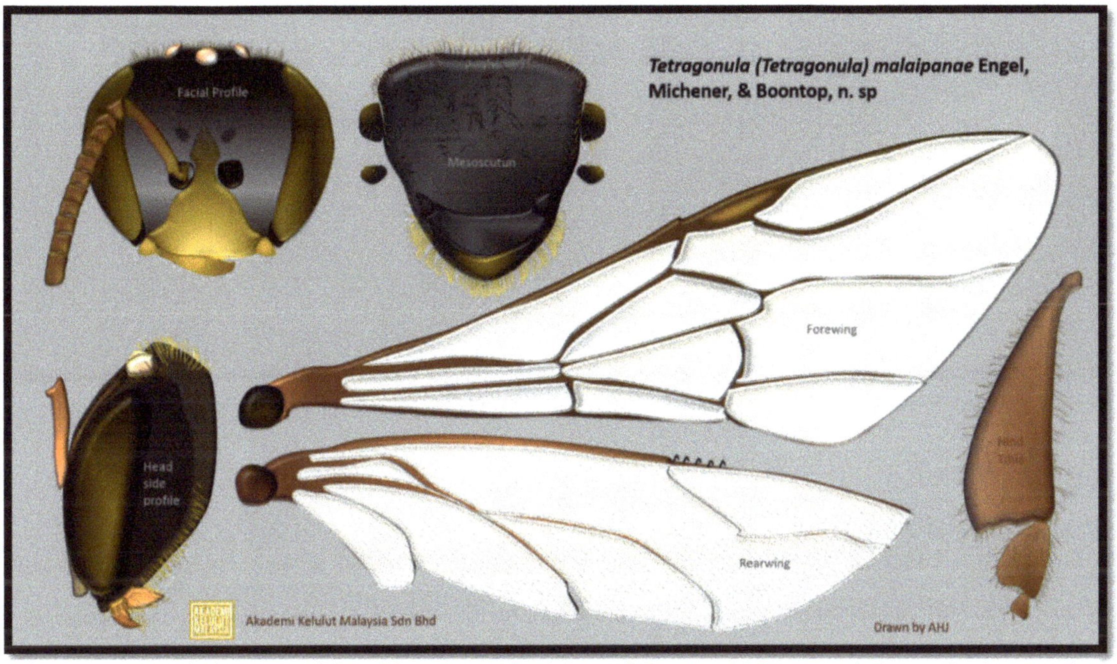

Chart 5 Tetragonula (Tetragonula) malaipanae Engel, Michener & Boontop 2017 anatomical parts & wing venation chart

***13.9.3 satun**, Lepidotrigona* Attasopa and Bänziger n. sp. 2018 Etymology. The specific epithet refers to the province in Thailand where the species was collected;

Map 2 Province of Sarun near the Thailand – Malaysia border

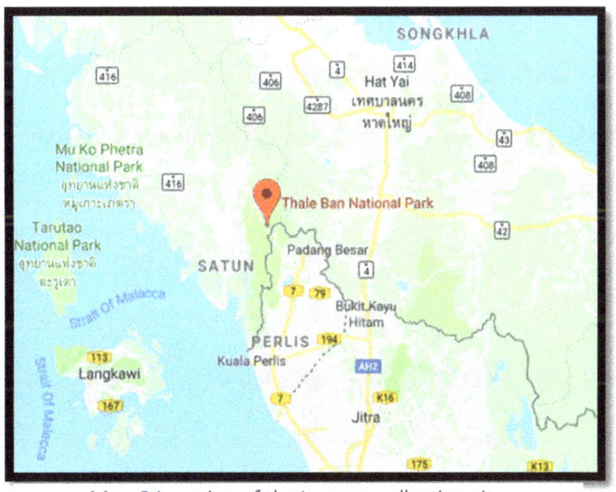

Map 3 Location of the L. satun collection site.

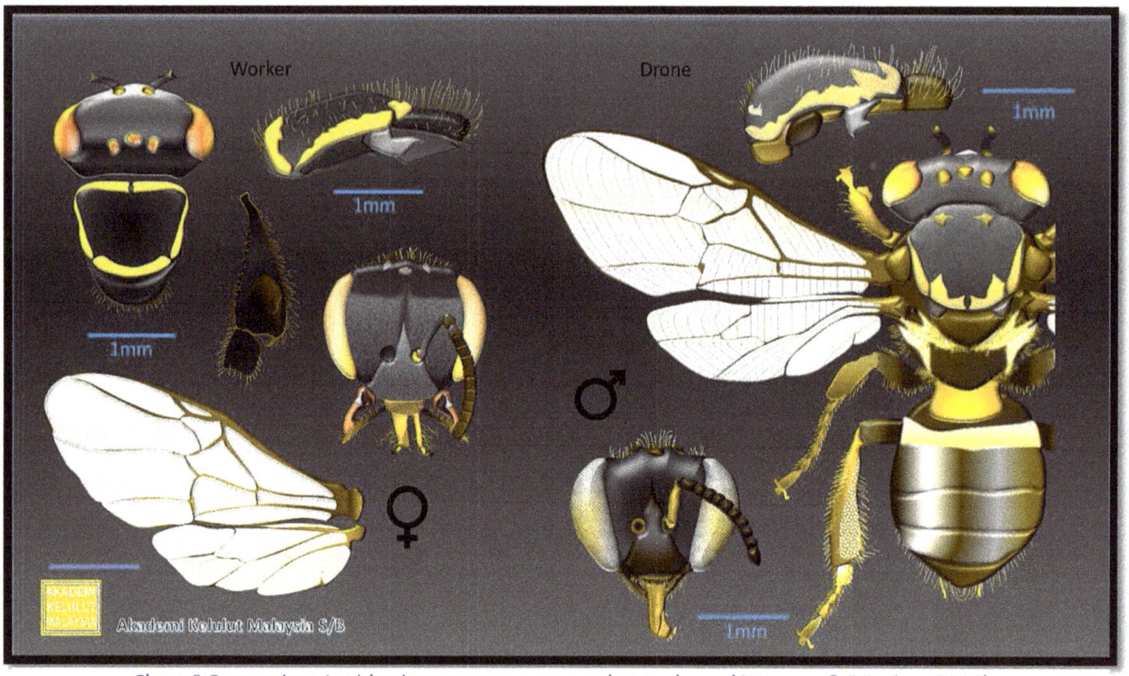

Chart 6 Comparison Lepidotrigona satun n. sp. worker to drone (Attasopa & Bänziger 2018)

Reference:

Engel M.S., Michener C.D., & Boontop Y., Notes on Southeast Asian Stingless Bees of the Genus Tetragonula (Hymenoptera: Apidae), Description of a New Species from Thailand. American Museum Novitates Number 3886, 17 pp. December 6, 2017.

Attasopa K., Bänziger H., Disayathanoowat T., & Packer L., A new species of Lepidotrigona (Hymenoptera: Apidae) from Thailand with the description of males of L. flavibasis and L. doipaensis and comments on asymmetrical genitalia in bees. Zootaxa 4442 (1): 063–082, 2018.

13.9.4 basimaculata, *Melipona* Bingham 1903* (Moure's Bee Catalogue, 2008) and is newly accepted in 2019 as *Tetragonula basimaculata* (Bingham, 1903) [TSN: 1129864 ITIS Bees Checklist] Etym: **basi** - Latin- **basis** - A pedestal, foot, base; + maculate (Latin –"spotted or stained"); having stained base or feet.

Notes: Biserat, Telom, Cameron Highlands. January 15, ***1902.*** Resting on the ceiling at midday. 'This species resembles *M.* ***ventralis*** a little but is very much larger and with different sculptures. (Bingham 1903)

Chart 7 Redrawn from silhouette illustration in Bingham 1903.

13.9.5 tricholoma, *Heterotrigona (Sahulotrigona)* Engel, new species [Etym.: The specific epithet is taken from the Greek words, trichós (τρῐχός), meaning "hair", and lôma (λῶμᾰ), meaning, "border" or "fringe", and is a reference to the fringe of setae bordering the mesoscutum posterolaterally.]

13.9.6 taraxis, *Heterotrigona (Sahulotrigona)* Engel, new species [Etym.: The specific epithet is taken from the Greek, *taraxis* (τᾰρᾰχῆς), meaning, "confusion", and is a reference to the historical confusion regarding the identity of this species.]

Comments: This species has been long confused as "*Trigona atricornis* Smith" a recent examination of the holotype of *T. atricornis* in the Oxford University Museum of Natural History reveals that Smith's (1865) species in actuality belongs to the genus *Papuatrigona* Michener & Sakagami, wherein it is a senior synonym of *T. genalis* Friese (*vide infra*) (Engel 2019).

Reference:

Michael S. Engel, Notes on Papuasian and Malesian stingless bees, with the descriptions of new taxa (Hymenoptera: Apidae) Journal of Melittology No. 88, pp. 1–25. August 12 2019.

13.10 Indian Subcontinent

13.10.1 Flavotetragonula, Etym: The subgeneric name is derived from flavus in Latin for yellow.

[1] **Abstract: A** new subgenus of stingless bees, *Flavotetragonula* Shanas, subgen. n. is established, and three new species, *Tetragonula (Flavotetragonula) calophyllae* Shanas and Faseeh, n. sp., *Tetragonula (Tetragonula) perlucipinnae* Faseeh and Shanas, n. sp. and *Tetragonula (Tetragonula) travancorica* Shanas and Faseeh, n. sp. is described from southern India, based on workers. Association for Advancement of Entomology 2019, DOI: 10.33307/entomon.v44i1.424

Remarks: *Flavotetragonula* Shanas is a new synonym of *Tetragonula* Moure s.str. (New synonymy), (Engel 2019).

13.10.2 calophyllae - *Tetragonula (Flavotetragonula) calophyllae* Shanas and Faseeh, n. sp. [1],

Etym: The specific epithet is after the generic name of the endangered tree *Calophyllum inophyllum*, on which the first feral colony was observed.

13.10.3 perlucipinnae - *Tetragonula (Tetragonula) perlucipinnae* Faseeh and Shanas, n. sp. [1]

Etym: The specific epithet is based on the Latin words perlucidulus and pinna, alludes to the transparent wing.

13.10.4 travancorica -*Tetragonula (Tetragonula) travancorica* Shanas and Faseeh, n. sp. [1],

Etym: This species is named after the erstwhile Kingdom of Travancore.

Remarks: The Kingdom of Travancore (/ˈtrævənkɔːr/) (Thiruvithamkoor) was an Indian kingdom from c. 870 until 1949. It was ruled by the Travancore Royal Family from Padmanabhapuram and later Thiruvananthapuram. The kingdom covered most modern-day central and southern Kerala at its zenith, with the Thachudaya Kaimal's enclave of Irinjalakuda Koodalmanikyam temple neighbouring Kingdom of Cochin, as well as the district of Kanyakumari, now in the Indian state of Tamil Nadu.

13.10.5 chandrai, *Lisotrigona chandrai* Viraktamath and Sajan Jose 2017[2] [syn. to *Lisotrigona cacciae* (Nurse 1907)] (Rasmussen 2017) [Etym: Dedicated to Indian taxonomist Dr Chandrashekhar A Viraktamath]

Remarks: This is named after a well-known Indian Taxonomist Dr Chandrashekhar A Viraktamath, popularly known as "Chandra" among his international colleagues and friends. Dr CAV (that is how he is known among his Indian students and colleagues) has dedicated his life to studying Indian insects, especially leafhoppers. He is a mentor and guides a host of Indian and International students, including the authors of this paper. Shortly after turning 75 and suffering from failing eyesight, he still works for 18 hours a day in his laboratory and home. Hence, this new species of stingless bees is named as a tribute to this great Indian personality. (see **Note)

13.10.6 revanai, *Lisotrigona revanai* Viraktamath and Sajan Jose 2017[2] [Etym: The species is named after a famous Hindu God, "Revanasiddeshwara". syn. to *Lisotrigona cacciae* (Nurse 1907)] (Rasmussen 2017). Type Loc. It is collected in Sawantawadi, Maharashtra, India. (Pers. Comm. Viraktamath, 2017)]

Remarks: The species is named after a famous Hindu God, "Revanasiddeshwara". The temple is located near the type locality of the new species on a hillock known as "Revanadi" in Maharashtra. Also known as a **Theonym** - A **theonym** (from classical Greek *theos* "god" and *-onym* "name") is a proper name of a deity.

13.10.7 kyrdemkulaiensis, *Tetragonula kyrdemkulaiensis* Viraktamath and Rojeet sp. n. [3] Etym: This species is named after the place where it is collected. Holotype: Male: Meghalaya: Kyrdemkulai (25.44° N, 91.47° E, Altitude 679 m a.s.l.), 20. ii. 2019, Coll. Rojeet T. deposited at UASB.

Remarks: This stingless bee was found in a subterranean nest at Kyrdemkulai, Ri-Bhoi District of Meghalaya, in January 2019. The nest was located at a depth of 90 cm in the soil with an entrance burrow (116 cm in length) running horizontally.

13.10.8 srikantanathi, *Tetragonula srikantanathi* Viraktamath sp. n. [3] This species is named in honour of Mr. Srikanta Nath. Holotype: Female: Tripura: Salema (24.01° N 91.83° E, Altitude 89 m a.s.l.), 8. xi. 2017, Coll. S. Viraktamath, deposited at UASB.

Remarks: Mr. Srikanta Nath, Assistant Director of Agriculture of the Tripura State Agriculture Department, searched the nest and helped study its structure.

Reference for New Names up to January 2020:

(1) Shanas S. & Faseeh P., A new subgenus and three new species of stingless bees Hymenoptera: Apidae: Apinae: Meliponini) from India. Entonom Vol 44 No 1 (2019)

(2) Viraktamath S. & Jose K. S., Two New Species of Lisotrigona Moure (Hymenoptera: Apidae: Meliponini) From India with Notes on Nest Structure. The Bioscan 12(1): 21-28, 2017

(3) Viraktamath S. & Rojeet T., Two new species of Tetragonula (Hymenoptera: Apidae: Meliponini) from North-East India with notes on their nest structure. Institute of Zoology, Slovak Academy of Sciences, January 20 2020

References for the History Section.

1) Hirst, K. K. (May 2016). Ancient Maya Beekeeping: Stingless Bee *Melipona beecheii (Ancient Domestication of the American Stingless Bee)*. Retrieved from About Education: http://archaeology.about.com/od/Domesticated-Animals/fl/Ancient-Maya-Beekeeping-The-Stingless-Bee-Melipona-beecheii.htm
2) Imre, D. M., Young, L., & Marcus, J. (n.d.). Ancient Maya Beekeeping (ca. 1000-1520 CE). *University of Michigan Undergraduate Research Journal*.
3) Roubik, D. W. (June 16 2005). Mayan Stingless Bee Keeping: Going, Going, Gone? *Smithsonian Tropical Research Institute*. Retrieved from https://www.sciencedaily.com/releases/2005/06/050615062105.htm
4) Wikipedia. (September 28 2016). *Beekeeping*. Retrieved from Wikipedia: https://en.wikipedia.org/wiki/Beekeeping
5) Wilson-Rich, N., Allin, K., Carreck, N., Quigley, A., The Bee a Natural History, *Ivy Press* 2014
6) G. Quezada-Euán, J.J. The Past, Present, and Future of Meliponiculture in Mexico; Stingless Bees of Mexico pp 243-269, *Springer Link*. (First Online: 04 August 2018)
7) G. Quezada-Euán, J.J., et al., Meliponiculture in Mexico: problems and perspective for development; Bee World 82(4): 160–167 (2001), *IBRA*

Works Cited in Part 3

Anne Dollin, Pers. Comm. (2015).

Bingham, C. (1903). Diagnoses Of Aculeate Hymenoptera. *Fasciculi Malayenses.*

Bold Systems. (2014). *taxid=488448.* Retrieved 19 11, 2014, from http://www.boldsystems.org/index.php/Taxbrowser_Taxonpage?taxid=488448

Boongird, S. &. Michener, C. D.. (2004). A New Species of *Trigona* from Peninsular Thailand (Hymenoptera: Apidae: Meliponini).April 2004 *JKES* 77(2):143-146

Claus Rasmussen, Pers. Comm.. (2013).

Creative commons. (SA 3.0 AU). *Attribution-ShareAlike 3.0 Australia.* Retrieved 11 11, 2014, from https://creativecommons.org/licenses/by-sa/3.0/au/

Discover Life. (17 11, 2014). *Tetragonula sarawakensis.* Retrieved 18 11, 2014, from http://www.discoverlife.org/mp/20q?search=Tetragonula+sarawakensis

Discover Life. (2014). *Tetrigona apicalis.* Retrieved 17 11, 2014, from http://www.discoverlife.org/mp/20q?search=Tetrigona+apicalis

Dollin et al., A. (1997). Australian Stingless Bees of the Genus *Trigona* (Hymenoptera: Apidae). *Invertebrate Taxonomy* 11(6) 861 - 896 *Csiro Publishing.*

Dollin, A. E., Dollin, L. J., & Rasmussen, C. (2015). A revision is the Australian and New Guinean Stingless Bees of the Genus Austroplebeia Moure (Hymenoptera: Apidae). — a revision, *Zootaxa* 4047 (1), pp. 1-73: 25

Doug Yanega. (September 31, 2007). *Bee Genera Of The World based on Michener, 2007.* Retrieved from http://cache.ucr.edu/~heraty/beepage.html

Encyclopedia of Life. (n.d.). *Trigona (Tetragona) fusco-balteata var. pagdeni Schwarz 1939, H. Smith (coll.), Nakon Sritamarat Siam, Thailand.* Retrieved 19 11, 2014, from H. Smith (coll.), Nakon Sritamarat Siam, Thailand: http://eol.org/pages/23270828/details

Encyclopedia of Life. (n.d.). *Trigona atomella.* Retrieved 16 11, 2014, from http://eol.org/pages/23268106/overview

Engel. (2005). A primitive sapygid wasp in Burmese amber (Hymenoptera: Sapygidae). *Acta zoologica cracoviensia*, 48.

Engel, M. S. (2000). A Review of the Indo-Malayan Meliponine Genus *Lisotrigona* with Two New Species (Hymenoptera, Apidae). AMNH *Oriental Insects*, 34:1, 229-237, DOI: 10.1080/00305316.2000.10417261

Engel, M. S. (2000). A New Interpretation of the Oldest Fossil Bee (Hymenoptera: Apidae) Number 3296, 11 pp., April 25, 2000 *American Museum Novitates.*

Engel, M. S. (2001). A Monograph Of The Baltic Amber Bees And Evolution Of The Apoidea (Hymenoptera). *Bulletin of The American Museum of Natural History.*

Engel, M. S., & Rasmussen, C. (2017). A new subgenus of Heterotrigona from New Guinea (Hymenoptera: Apidae). *Journal of Melittology*.

Engel, M. S., Michener, C. D., & Boontop, Y. (2017). Notes on Southeast Asian Stingless Bees of the Genus Tetragonula (Hymenoptera: Apidae), with the Description of a New Species from Thailand. *American Museum Novitates*.

Gazi, T. (July 2013). Collections At The Museum: The Pendlebury Collection. *Bulletin of the Museum of Zoology Volume 1 Issue 3*, p. 5.

Goulet, H., & Huber(eds), J. T. (1993). Hymenoptera of the world: An Identification Guide to Families. *Agriculture Canada*.

Jalil, A. (2014). *Beescape for Meliponines.* Singapore: Partridge Publishing.

Jalil, A. (2016). *Meliponine Identifier Pictoruial Guide.* Akademi Kelulut Malaysia.

Jalil, A. (2016). *World Meliponine Etymology of Taxanomic Nomenclature.* Akademi Kelulut Malaysia.

Jalil, A., & Roubik(ed), D. (2016). *Handbook of Meliponiculture.* Akademi Kelulut Malaysia.

Lepeletier, d. S. (1836). Des Insectes Hyménoptères. In A. L. Saint.Fargeau, *Histoire Naturelle* (P. 329). Paris, France: Î.Ibilairie Eïrcyci.Opédiqux De Roret,.

Michener, C. D. (2000). Bees Of The World. JHU Press.

Michener, C. D. (2007). The Bees of the World 2nd ed. *The John Hopkins university press*.

Michez, D., Vanderplanck, M., & Engel, M. S. (2012). Fossil bees and their plant associates. *The Systematics Association*.

Moure's Bee Catalogue. (23 Jul, 2008). *Bee Catalogue*. (Universidade Federal do Paraná, UFPR, Centro de Referência em Informação Ambiental, CRIA) Retrieved 17 Jun, 2013, from Moure's Bee Catalogue: http://moure.cria.org.br/catalogue

Nogueira-Neto, P. (1997). Life Creating Bees.

Rahman, A., Das, P. K., Rajkumari, P., Saikia, J., & Sharmah, D. (2015). Stingless Bees (Hymenoptera:Apidae: Meliponini): Diversity and Distribution in India. *International Journal of Science and Research (IJSR)*, Volume 4 Issue 1.

Rasmussen. (2008). Catalogue of the Indo-Malayan/Australasian stingless bees (Hymenoptera: Apidae: Meliponini).

Rasmussen, C., Thomas, J. C., & Engel, M. S. (2017). A New Genus of Eastern Hemisphere Stingless Bees (Hymenoptera: Apidae), with a Key to the Supraspecific Groups of Indomalayan and Australasian Meliponini. *American Museum Novitates*.

Schwarz, H. F. (1937). Bornean Stingless Bees Of The Genus Trigona. *Bulletin American Museum of Natural History*.

Schwarz. (1938). Retrieved from http://www.gbif.org/species/1345534

Schwarz, H. F. (1939). The Indo Malayan Species of Trigona. *Bulletin of AMNH*.

Senckenberg Biographies. (December 11, 2013). *World of Biodiversity*, 6. (Senckenberg) Retrieved April 20, 2014, from Biographies of entomologists of the world: http://sdei.senckenberg.de/biographies/

Viraktamath, S., & Jose, K. S. (2017). Two New Species Of Lisotrigona Moure (Hymenoptera:Apidae: Meliponini) From India With Notes On Nest Structure. *The Bioscan 12(1): 21-28, 2017*, 21-28.

Winston. (1999). Practical Taxonomic Procedure for Biologists. In *Describing Species.*

Xiong, C. S. (2015). A Conservation Assessment of Stingless Bees (Apidae: Meliponini) in Singapore. *Cohort AY2014/2015 S1*.

Web references

Wikipedia. (July 20, 2013). *Frederick Smith*. Retrieved April 21, 2014, from http://en.wikipedia.org/wiki/Frederick_Smith_(entomologist)

Wikipedia. (December 28, 2013). *Theodore Dru Alison Cockerell*. Retrieved April 20, 2014, from http://en.wikipedia.org/wiki/Theodore_Dru_Alison_Cockerell

Wikipedia. (February 7, 2014). *Adolpho Ducke*. Retrieved April 20, 2014, from http://en.wikipedia.org/wiki/Adolpho_Ducke

Wikipedia. (03 11, 2014). *Alfred Russel Wallace*. Retrieved 07 11, 2014, from http://en.wikipedia.org/wiki/Alfred_Russel_Wallace

Wikipedia. (28 10, 2014). *Charles Thomas Bingham*. Retrieved 07 11, 2014, from http://en.wikipedia.org/wiki/Charles_Thomas_Bingham

Wikipedia. (17 08, 2014). *E. C. Stuart Baker*. Retrieved 07 11, 2014, from http://en.wikipedia.org/wiki/E._C._Stuart_Baker

Wikipedia. (10 04, 2014). *Hugo Berthold von Buttel-Reepen*. Retrieved 07 11, 2014, from http://es.wikipedia.org/wiki/Hugo_Berthold_von_Buttel-Reepen

Wikipedia. (30 10, 2014). *Ida Laura Pfiefer*. Retrieved 7 11, 2014, from Ida_Laura_Pfeiffer: http://en.wikipedia.org/wiki/Ida_Laura_Pfeiffer

Wikipedia. (March 18, 2014). *Johann Karl Wilhelm Illiger*. Retrieved May 16, 2014, from http://en.wikipedia.org/wiki/Johann_Karl_Wilhelm_Illiger

Wikipedia. (March 15, 2014). *Louis Jurine*. Retrieved May 7, 2014, from http://en.wikipedia.org/wiki/Louis_Jurine: http://en.wikipedia.org/wiki/Louis_Jurine

Wikipedia. (03 09, 2014). *Mount Dulit*. Retrieved 16 11, 2014, from http://en.wikipedia.org/wiki/Mount_Dulit

Wikipedia. (05 09, 2014). *Mount Ophir*. Retrieved 16 11, 2014, from http://en.wikipedia.org/wiki/Mount_Ophir

Wikipedia. (28 10, 2014). *Proto-Malay*. Retrieved 20 11, 2014, from http://en.wikipedia.org/wiki/Proto-Malay

Part 4

The Dawn of *Tetragonula* Moure 1961

14.0 Prologue to Part 4

Nascency: The initial stage of a developmental process: beginning, birth, commencement, dawn, genesis, inception, nascence, onset, opening, origin, outset, spring, start. When considering the start of a new age, 'nascency' comes to mind. During the nascency of *Tetragonula* meliponiculture at the turn of the new millennium, *Tetragonula* bees were not a favourite choice for meliponiculture among stingless beekeepers in the ASEAN region. They were tiny, limited pollination range, short flight distances and poor honey producers. They could hardly come to par with their cousins of other meliponine genera like *Homotrigona*, *Geniotrigona* or *Heterotrigona* in terms of honey output. They are certainly set aside and never even got a second glance from Apiarists.

Compared to Apiculture, Meliponiculture is currently in its infancy. The interest in its honey's medicinal value grew, especially in areas that did not benefit from their larger 'cousins'. Namely, India, The Philippines, Java and Australia only had a better selection of *Tetragonula* to contend with. The fundamental quest for knowledge drives one to explore and discover all the peculiarities that set it apart from Apiculture. Essentially, one of the major differences in Apis bees' brood function and configuration is the variations of the meliponine brood structure.

This part 4 exhibits the diversity of the brood structures throughout Indo-Malaya and Australasia. The distribution of varied brood forms in the different regions corresponds to the species occurrence in the respective regions.

What is the citizen scientist, if not the child in us who never grows up, is ever curious and ever-inquisitive? This book is not necessarily for the ardent scientist but just spread data for the knowledge-seeking stingless beekeepers and enthusiasts. The manuscript is compiled during the Covid 19 Pandemic, and the data from the image bank is a compilation of more than a decade of fieldwork from the new millennium to date.

I understand that the adult bees need to be examined from the corresponding brood forms. However, I have some images of brood forms from Dr. Anchalee, and she has records that Michener identified. Some were based on brood descriptions from papers, like *T. drescheri* combination brood form description by Prof Siti Salmah, and it is unique and unmistakably easy to identify.

I address the *Flavotetragonula* issue with Engel's comments that it has been put in as a new synonym. He mentioned that they were named from examinations of teneral adults in a couple of them, and I have indicated this as some notes in Part 3. I also have those brood form images from the Indian authors who provided me with brood images of the ones they named in their journal publications. However, all these verifiable images (inc. *Austroplebeia*) constitute only 20% of the lot.

Still, Christoph Gutter has indicated in his book on "Stingless Bees Evolution and diversity" (SPRINGER) the following: *Tetragonula (Flavotetragonula)*

- *T. calophyllae* (Shanas & Faseeh, 2019)
- *T. gressitti* (Sakagami, 1978)
- *T. praeterita* (Walker, 1860)

Anyway, I have left some questions unanswered towards the end of part 2, especially regarding the evolution of the brooding form, whether the cluster evolved from the comb or vice-versa. I saw the passage below from Christoph's book.

Evolution of Nest Architecture

Variation in nest architecture often seems unrelated to stingless bee phylogeny, making it challenging to determine whether particular nesting traits are ancestral or derived. Several authors have suggested that clustered cells, which show similarities with the brood cell arrangement in bumblebees and some orchid bees, represent the ancestral state (Michener 1964; Camargo 1970; Wille and Michener 1973). This situation would imply that horizontal combs evolved from clustered cells. However, the phylogenetic relationships among stingless bee species suggest that repeatedly reversions from combs to clusters. Wille and Michener (1973) argued that the rather spherical cells that open in various directions are found in the cluster building Australian *Austroplebeia,* or some Neotropical *Trigonisca* represent the ancestral condition, whereas the oval and upward-oriented cells in clusters of *Frieseomelitta* spp. are derived. Indeed, the phylogenetic position of *Frieseomelitta* and Trichotrigona (Fig. 2.2) concerning their closest Neotropical relatives (mostly comb-building genera) suggests that, in their case, cell clusters are derived from brood combs (Christoph, 2020).

I was just wondering your take on the evolution of the brooding structure in Tetragonula specifically. In Tetragonula, the brood forms vary from cluster brood, semi-comb brood, spiral and layered combs. Most combs are horizontal while sometimes inclined or tilted regardless of cavity shape (although the cavity host could be inclined). But still, it does not confirm the evolution of the Tetragonula brood definitively.

I have added an appendix with charts depicting the locations of Meliponiculture activities and the meliponines that stingless beekeepers keep and breed. It gives an inkling of the type of bees that are commonly found and robust enough for propagation and product development. In some, they are kept purely for pollination, especially those on the highlands, because continuous honey harvesting may jeopardize their sustainability.

So, as you suggest, they remain hypothetical and suspect with all the uncertainties. I gear the writing towards a preliminary finding and require further study and verifications.

Chapter 14

Regional Distribution of *Tetragonula*

14.1 Section 1: The Indian Subcontinent

14.1.I Ellipsoidal Clustered Brood-cells

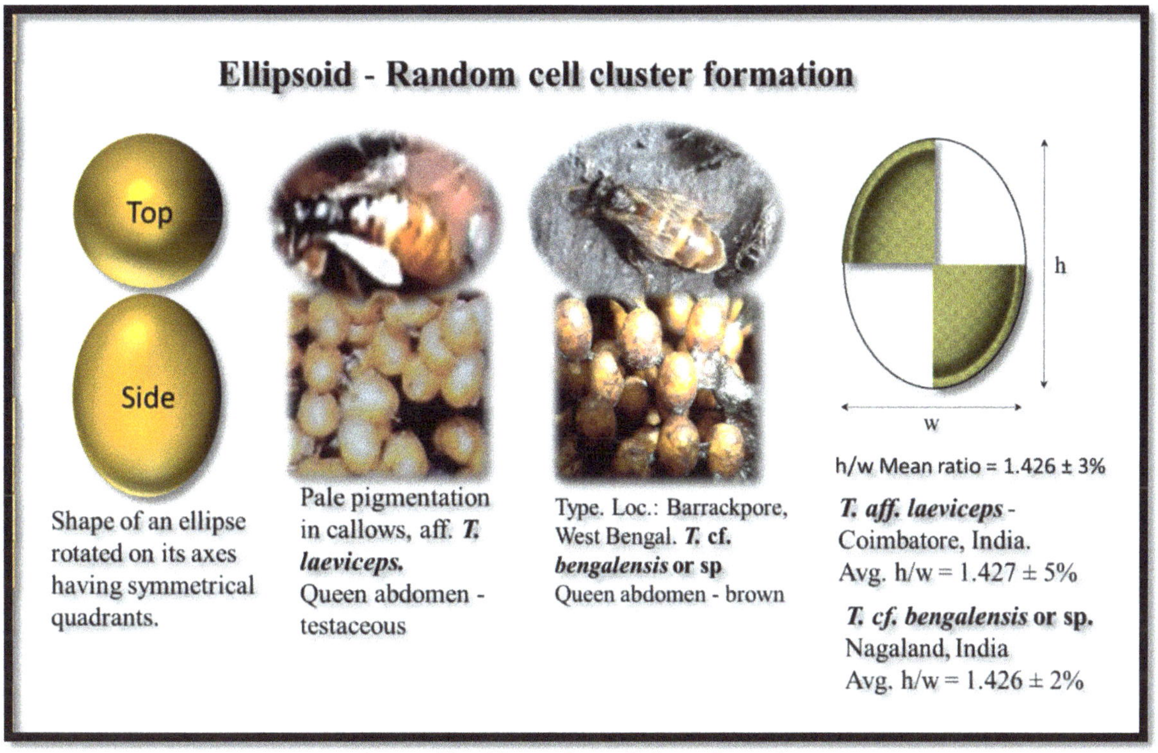

Chart 8 Ellipsoid - Random cell cluster formation

14.1.2a Ovaloid clustered cells

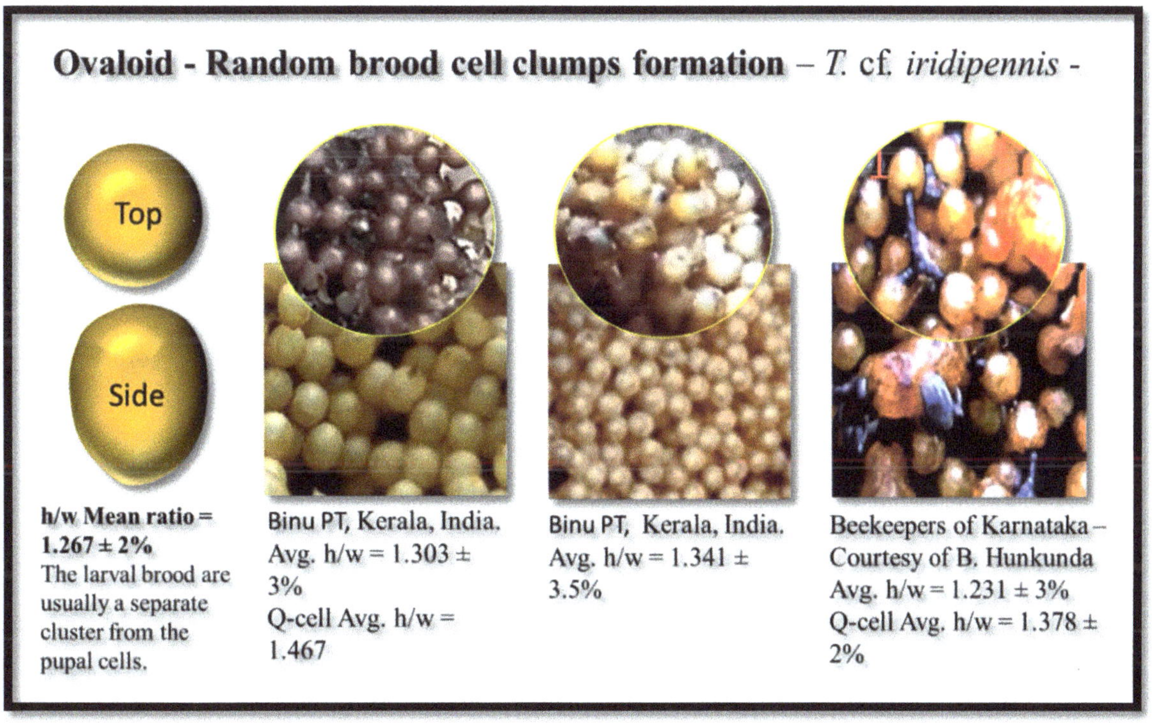

Chart 9 Ovaloid - Random brood cell clumps formation

14.1.2b Ovaloid clustered cells – off-white to cream cells

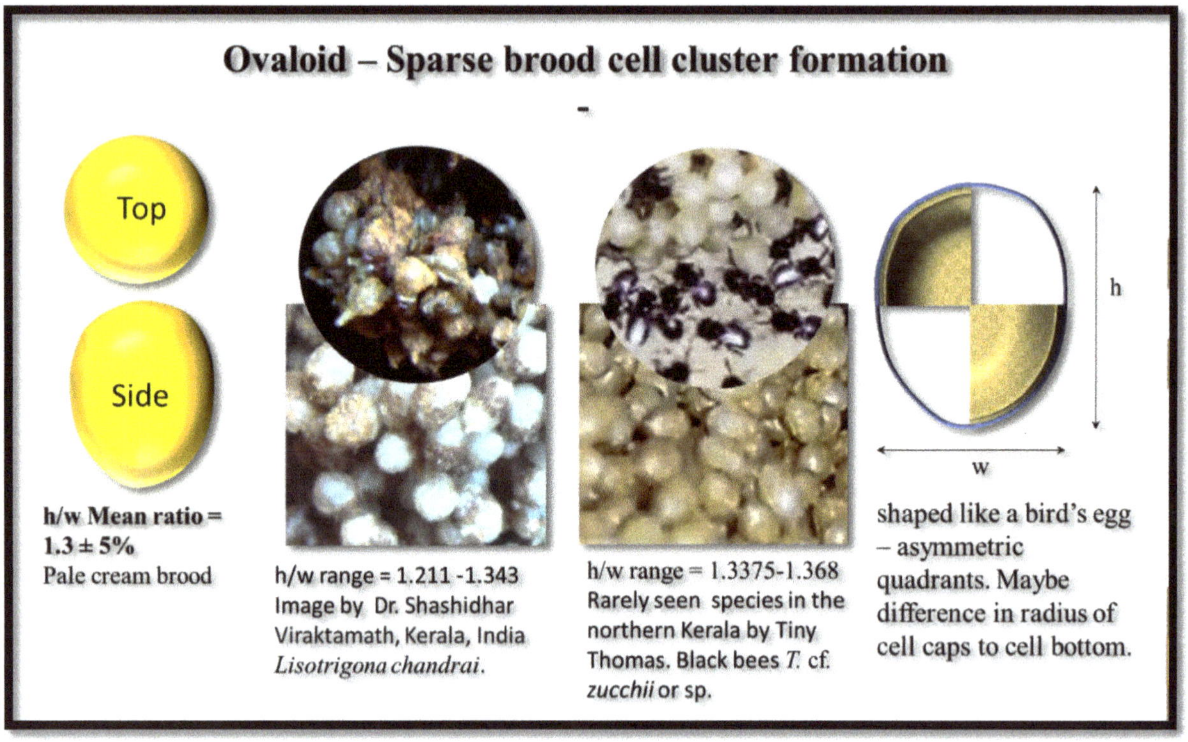

Chart 10 Off-white to cream coloured Prolate Spheroid cells

14.1.3 Cylindroid clustered cells

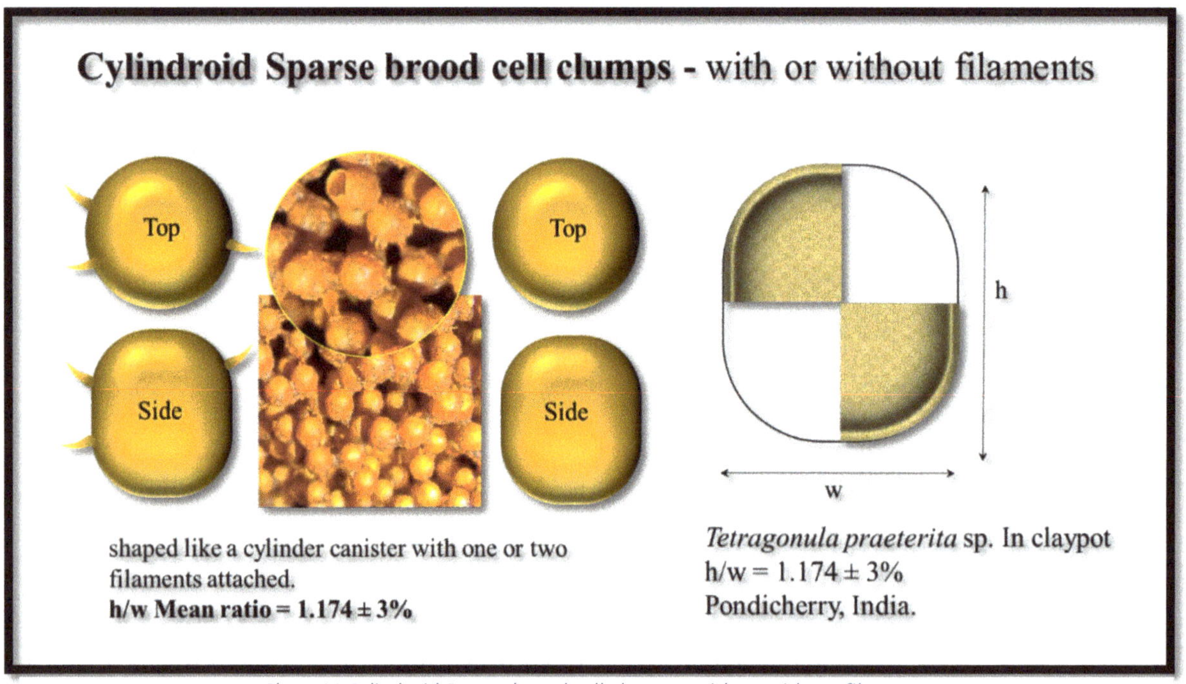

Chart 11 Cylindroid Sparse brood cell clumps - with or without filaments

14.1.4 Mixed Cylindroid & Ovaloid

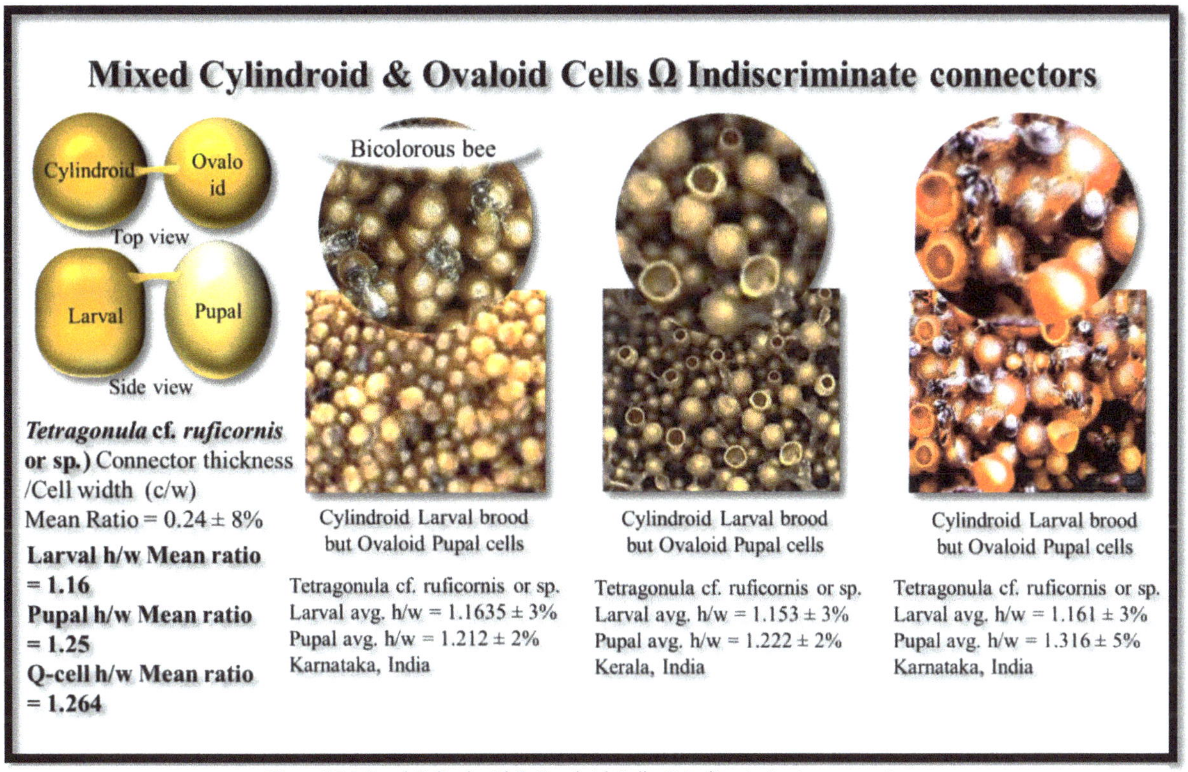

Chart 12 Mixed Cylindroid & Ovaloid Cells Ω Indiscriminate connectors

14.1.5 Cylindroid Convex layered

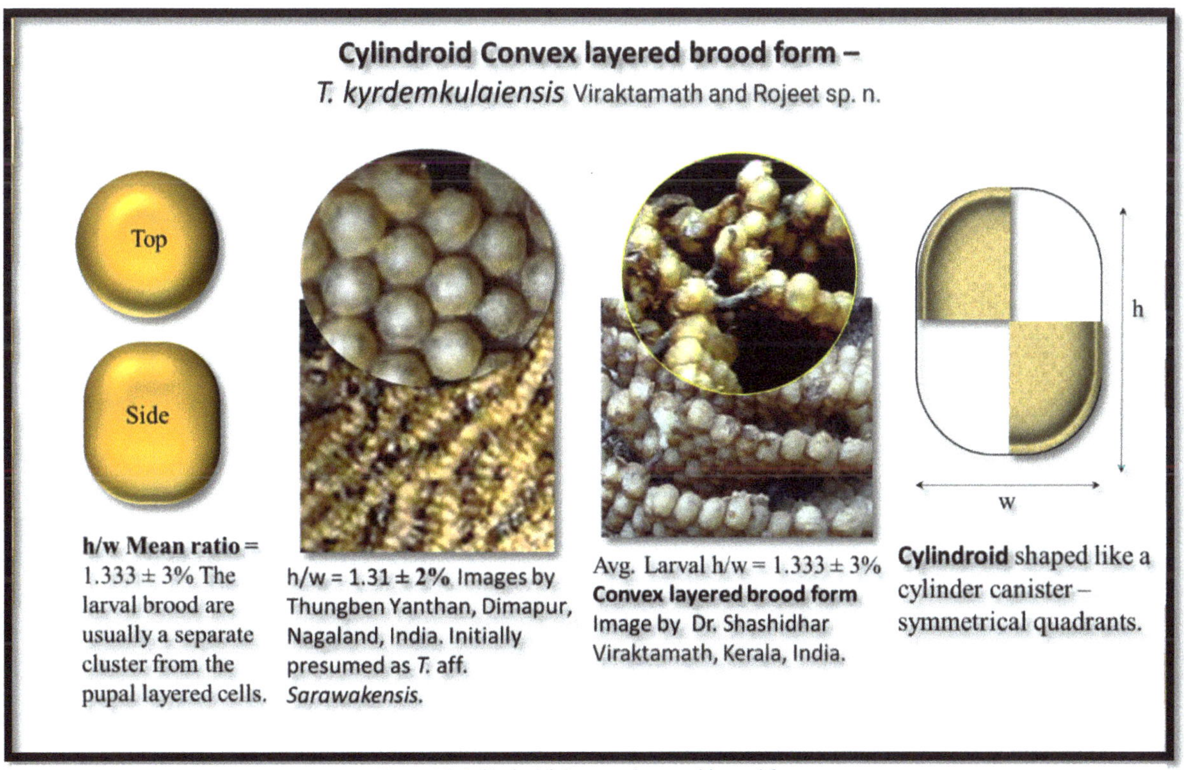

Chart 13 Cylindroid Convex layered brood form

14.1.6 Mixed Prolate Spheroid & Ovaloid

Mixed Prolate Spheroid Larval & Ovaloid Pre-pupal and Pupal Brood cells.

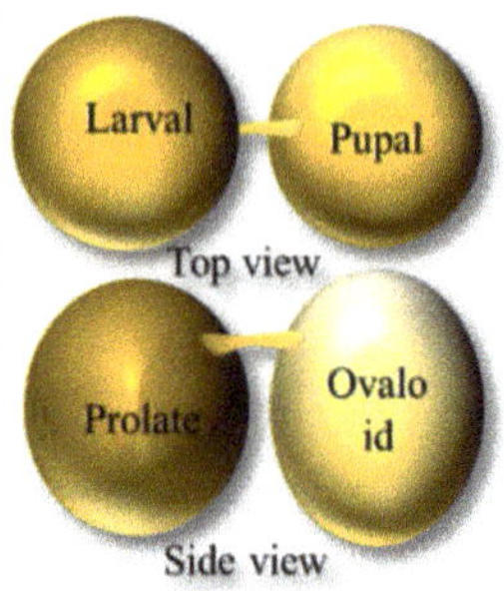

The larval brood are darker coloured and usually a separate cluster from the Pre-pupal and Pupal Brood cells.

Images by Dr. Shashidhar Viraktamath, Kerala, India. *T. srikantanathi* Viraktamath sp. n. Larval Avg. h/w = 1.167
Pre-pupal Avg. h/w = 1.36
Pupal Avg. h/w = 1.411 ± 2%

Images by Dr. Shanas S. *T. (Flavotetragonula) calophyllae* Shanas and Faseeh, n. sp.
Prolate Larval h/w = 1.132
Ovaloid Pupal h/w = 1.387

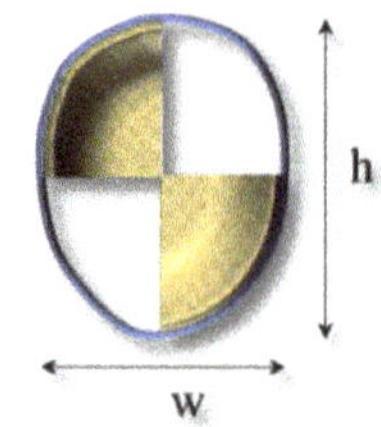

Ovaloid shaped

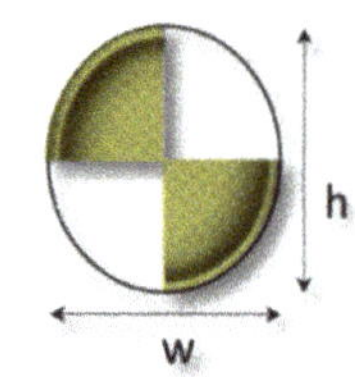

Prolate Spheroid

Chart 14 Mixed Prolate Spheroid & Ovaloid

14.1.7 Prolate Spheroid clusters & Ovaloid combs

Prolate Spheroid perforated clusters & Ovaloid combs

Tetragonula (Tetragonula) perlucipinnae Faseeh and Shanas, n. sp

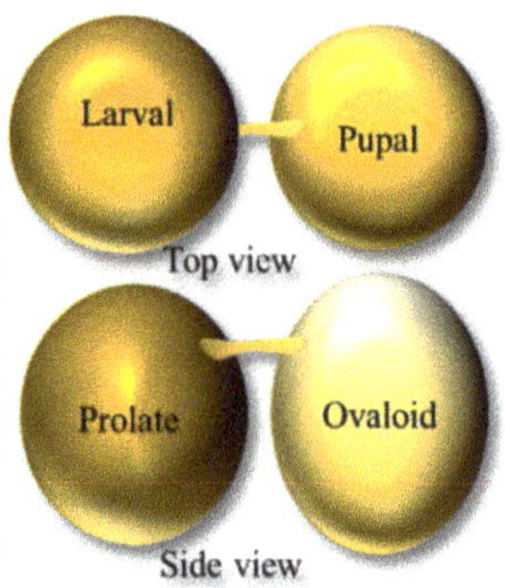

Prolate Larval Spheroid perforated clustered brood form h/w range = 1.12 -1.151;
Ovaloid pre-pupal perforated semi=comb brood form h/w = 1.229;
Ovaloid pupal full comb layer h/w = 1.18-1.294 brood can be taken apart in layers carefully.

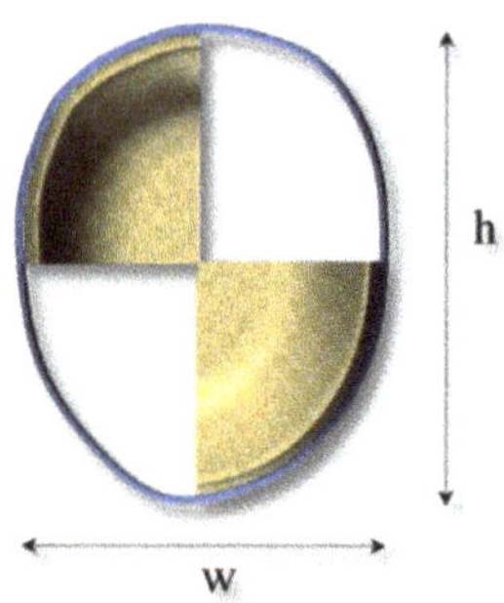

Ovaloid shaped like a bird's egg – asymmetric quadrants.
Pre-pupal Avg. h/w = 1.229
Pupal h/w range = 1.18-1.294

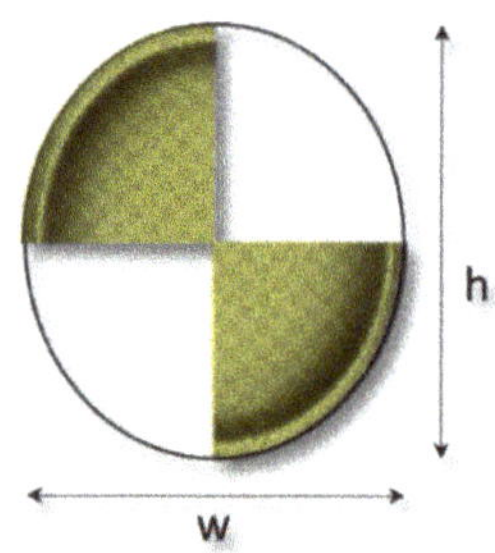

Prolate Spheroid Larval h/w range = 1.12 -1.151

Chart 15 Prolate Spheroid clusters & Ovaloid combs

14.1.8 Prolate Spheroid & Ovaloid clusters - graduated brood

Chart 16 Prolate Spheroid & Ovaloid clusters - graduated brood

14.1.9 Ovaloid Full Comb

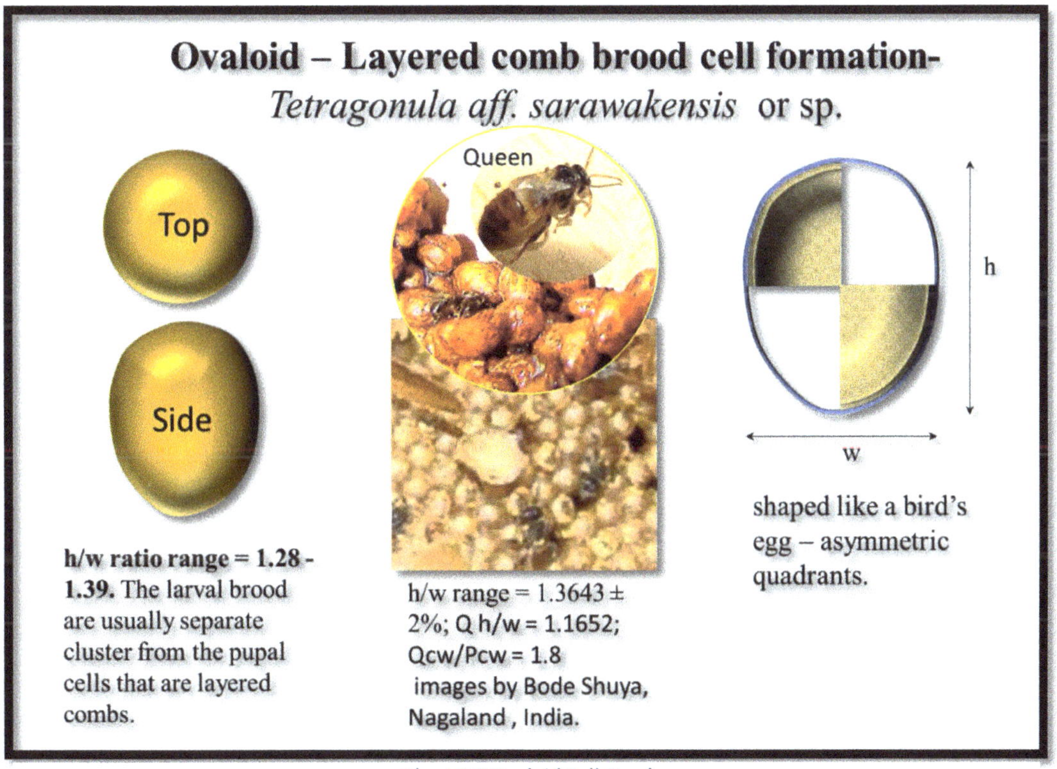

Chart 17 Ovaloid Full Comb

Section 2: Peninsula Malaysia. Thailand & Indo-China

14.2.1 Ellipsoidal Clustered Cells

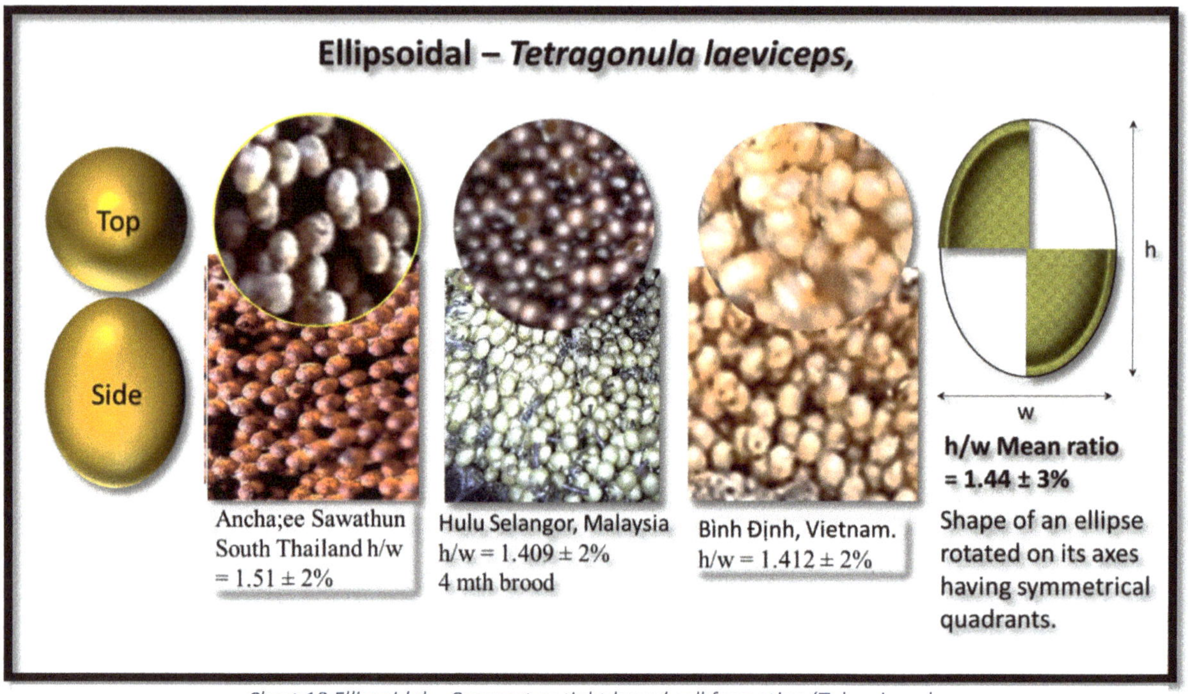

Chart 18 Ellipsoidal – Compact or tight brood cell formation (T. laeviceps)

14.2.2 Ellipsoid & Prolate Spheroid Cells

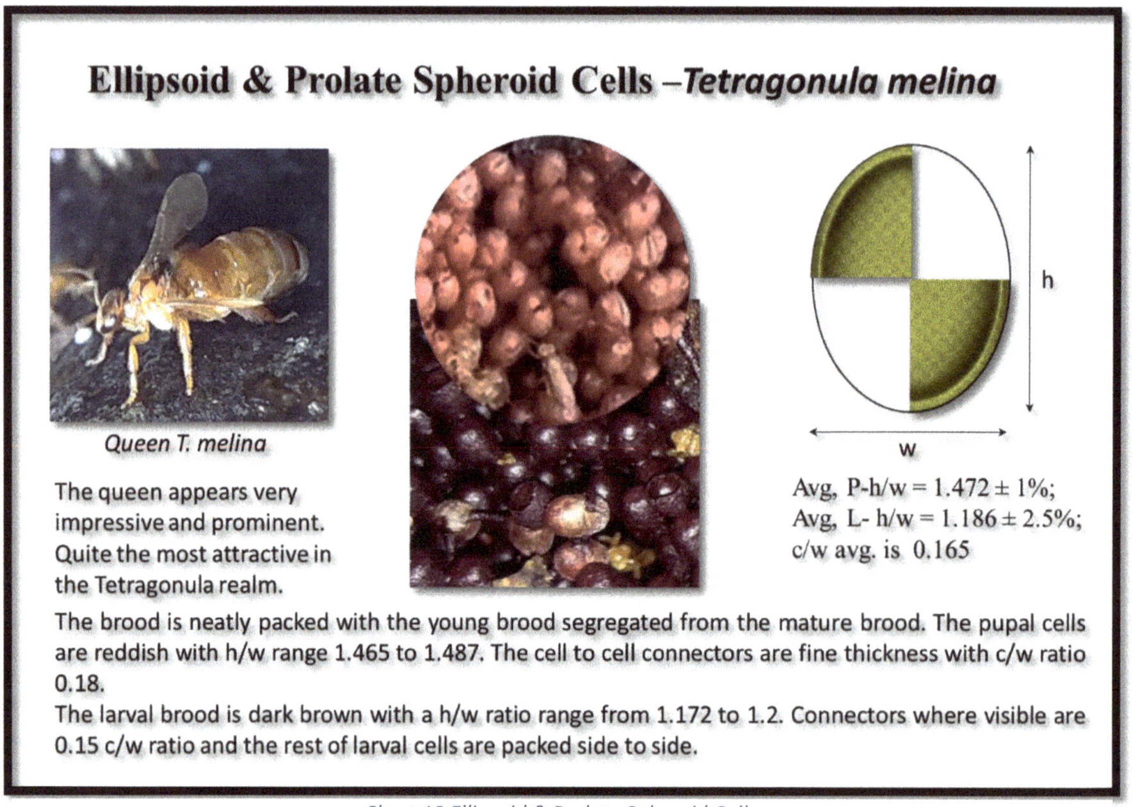

Chart 19 Ellipsoid & Prolate Spheroid Cells

14.2.3 Ovaloid – Perforated brood

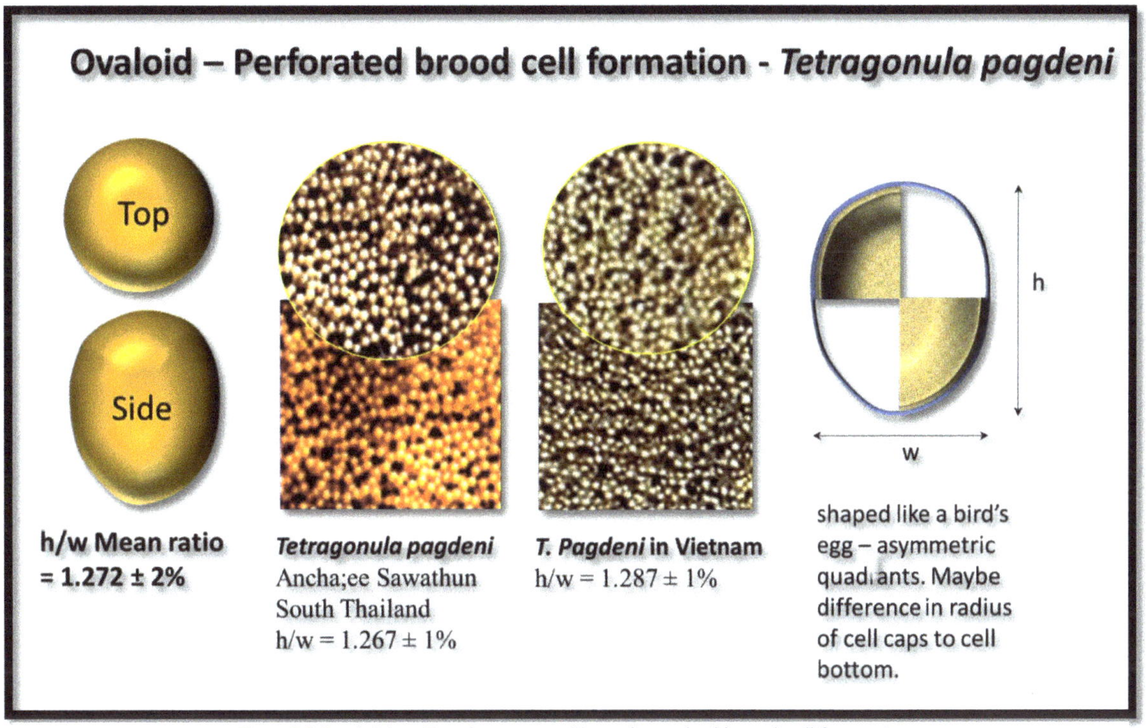

Chart 20 Ovaloid — Random spread loosely packed brood cell formation

14.2.3 Ovaloid – Clustered brood with thick connectors

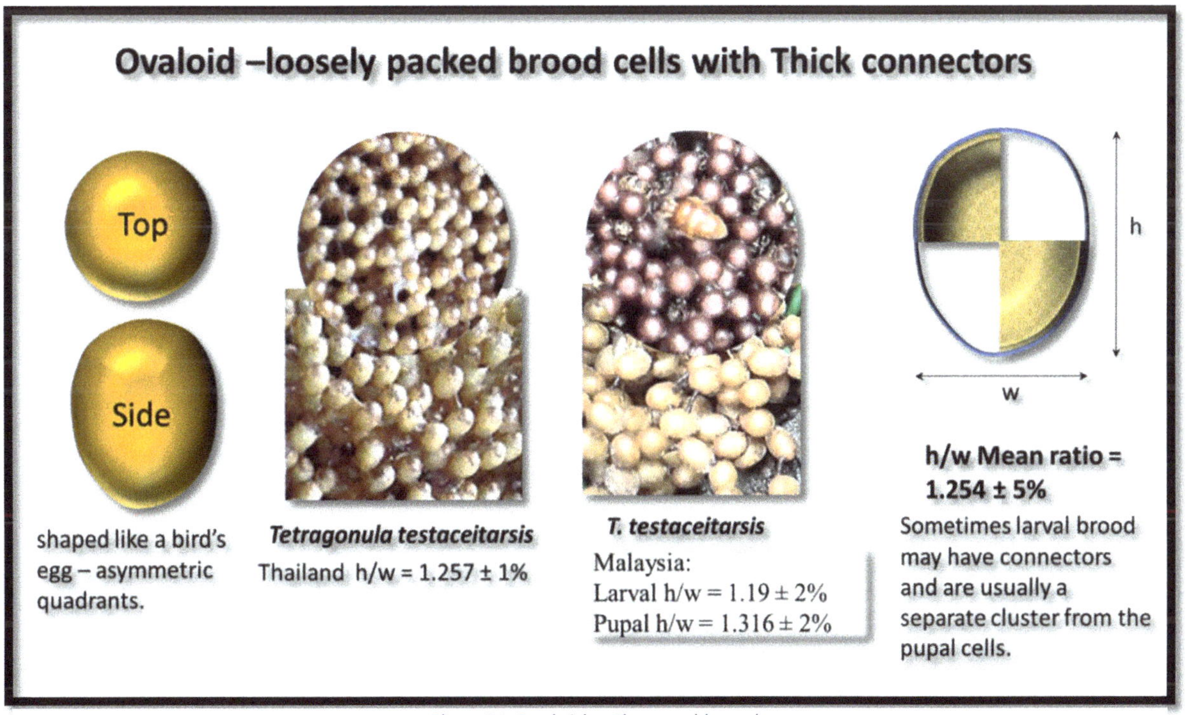

Chart 21 Ovaloid – Clustered brood

14.2.4 Oblate spheroid –globular brood cells

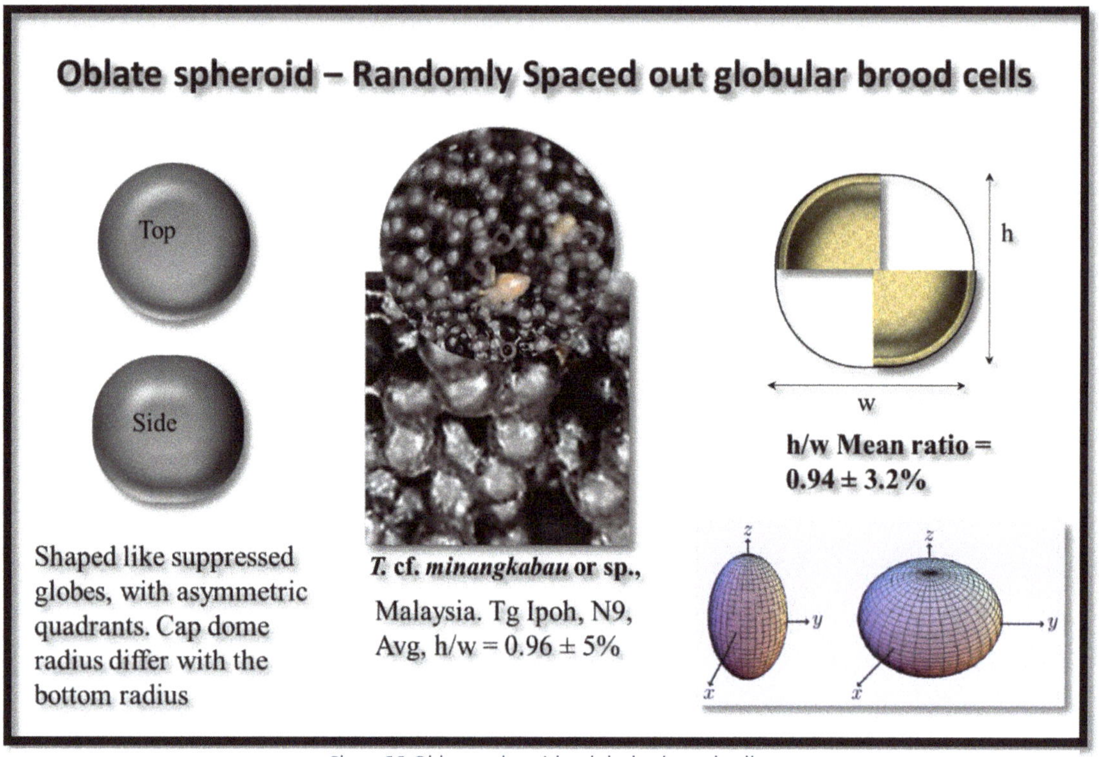

Chart 22 Oblate spheroid –globular brood cells

14.2.5 Mixed Spheroids & Ovaloid – semi-combs & clustered brood

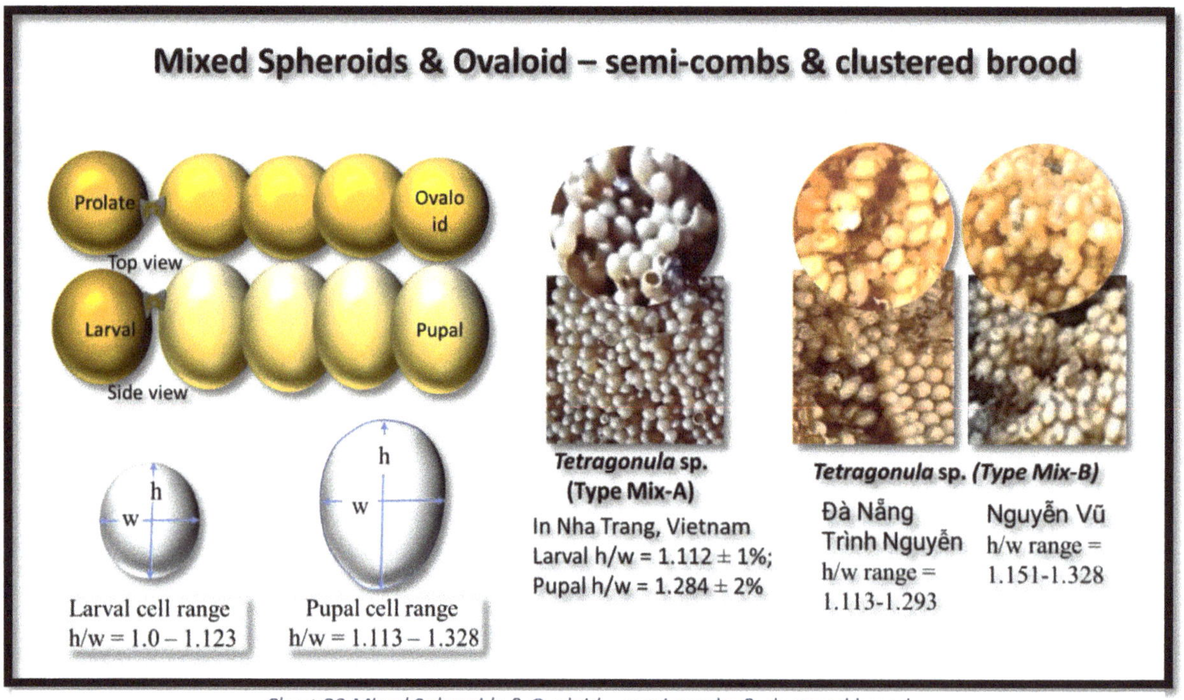

Chart 23 Mixed Spheroids & Ovaloid – semi-combs & clustered brood

14.2.6 Ellipsoidal & Ovaloid – Semi comb & cluster

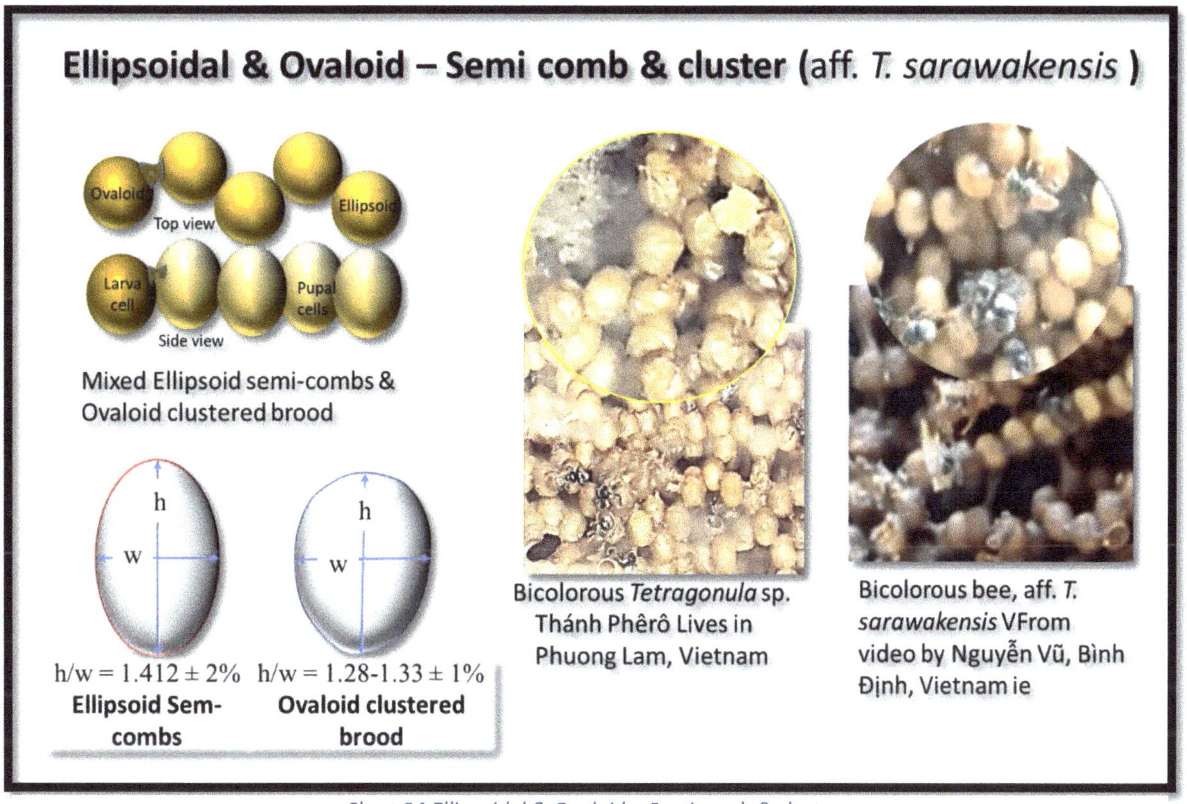

Chart 24 Ellipsoidal & Ovaloid – Semi comb & cluster

14.2.7 Prolate spheroid – Clustered clumps

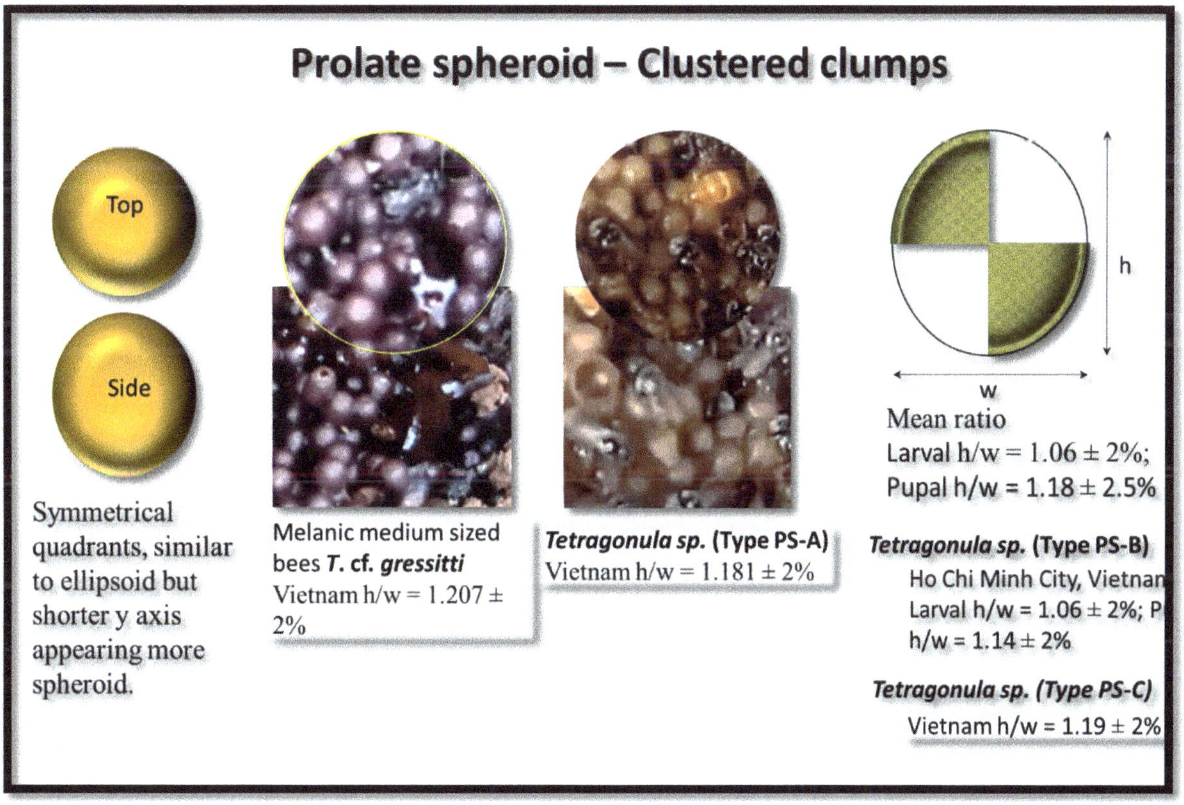

Chart 25 Prolate spheroid – Clustered clumps

14.2.8 Mixed Prolate Spheroid & Ovaloid Cells

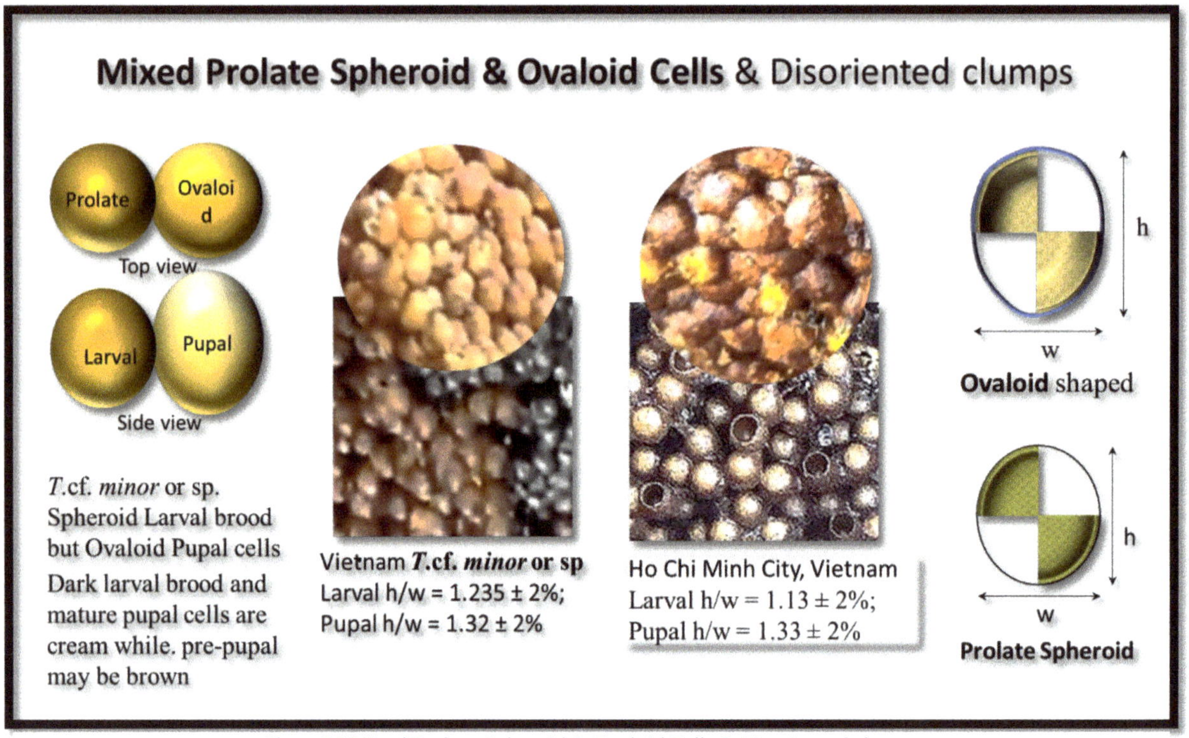

Chart 26 Mixed Prolate Spheroid & Ovaloid Cells & Disoriented clumps

14.2.9 Mixed Ellipsoid, Ovaloid and Cylindroid

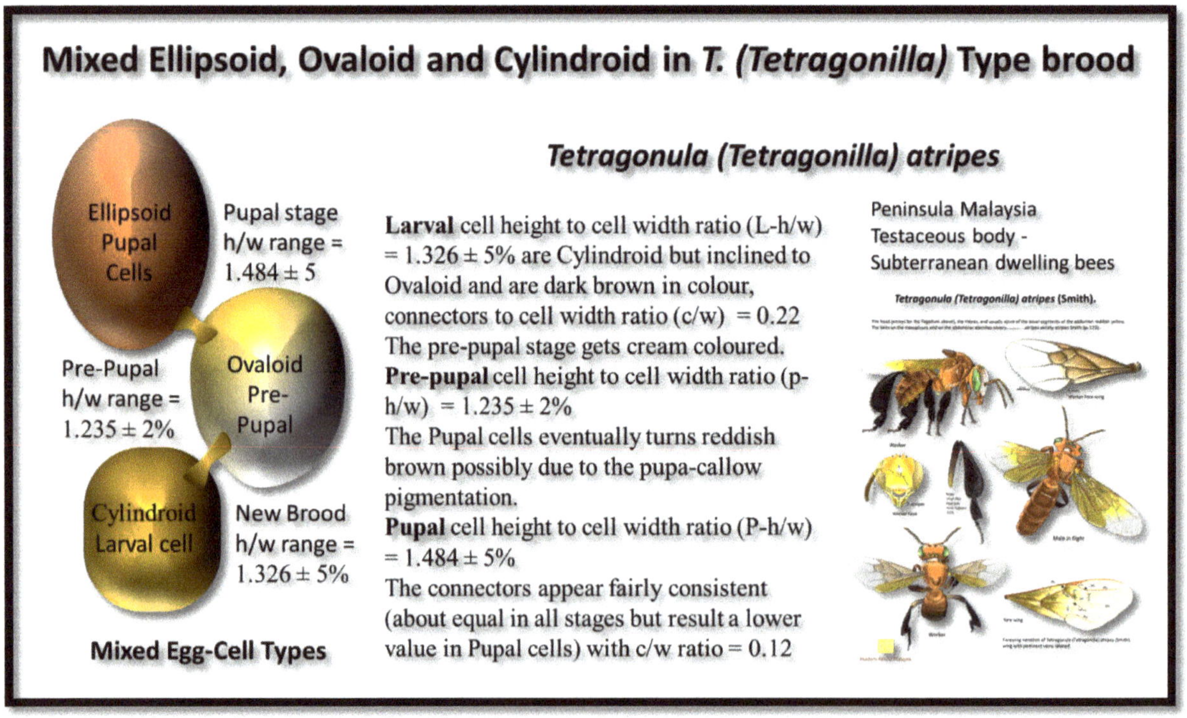

Chart 27 Mixed Ellipsoid, Ovaloid and Cylindroid in T. (Tetragonilla) Type brood

14.2.10 Mixed All Shapes

Chart 28 Ellipsoid, Ovaloid and Spheroid – loose brood structured Tetragonula (Tetragonilla) collina,

Section 3: Greater Sunda Islands & Java

14.3.1a Ellipsoidal - Compact or tight brood

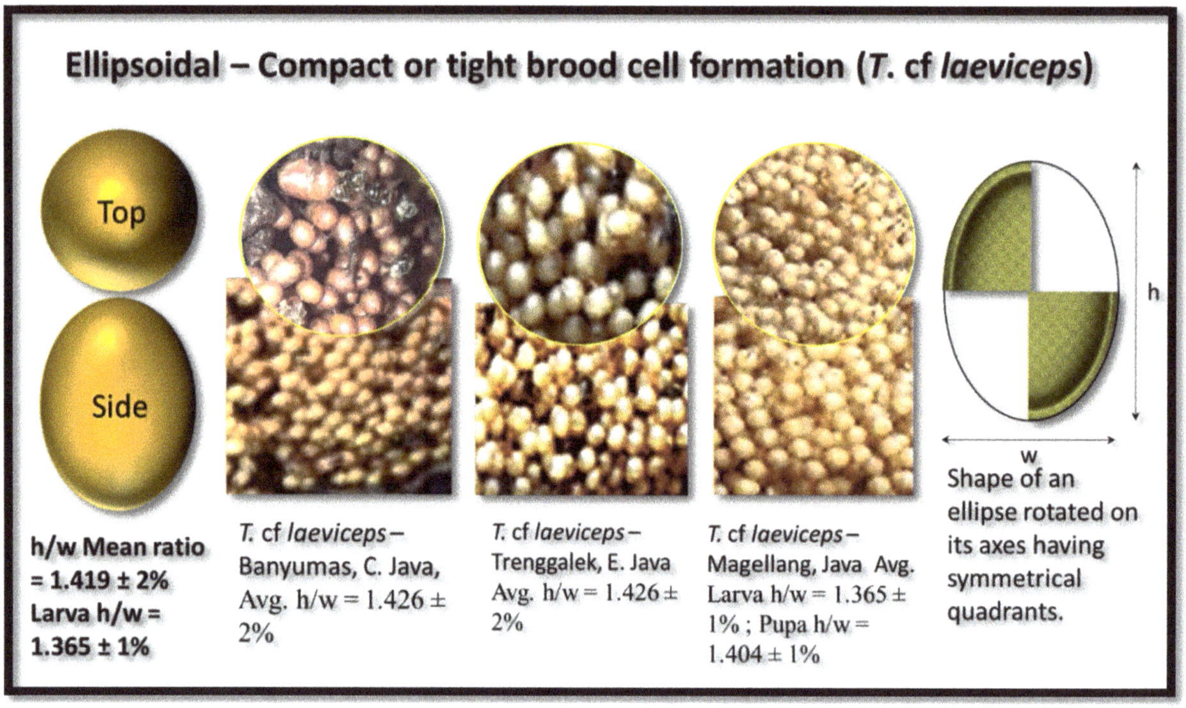

Chart 29 Ellipsoidal – Compact or tight brood cell formation (T. cf laeviceps) Borneo

14.3.1b Ellipsoidal – gravelly brood

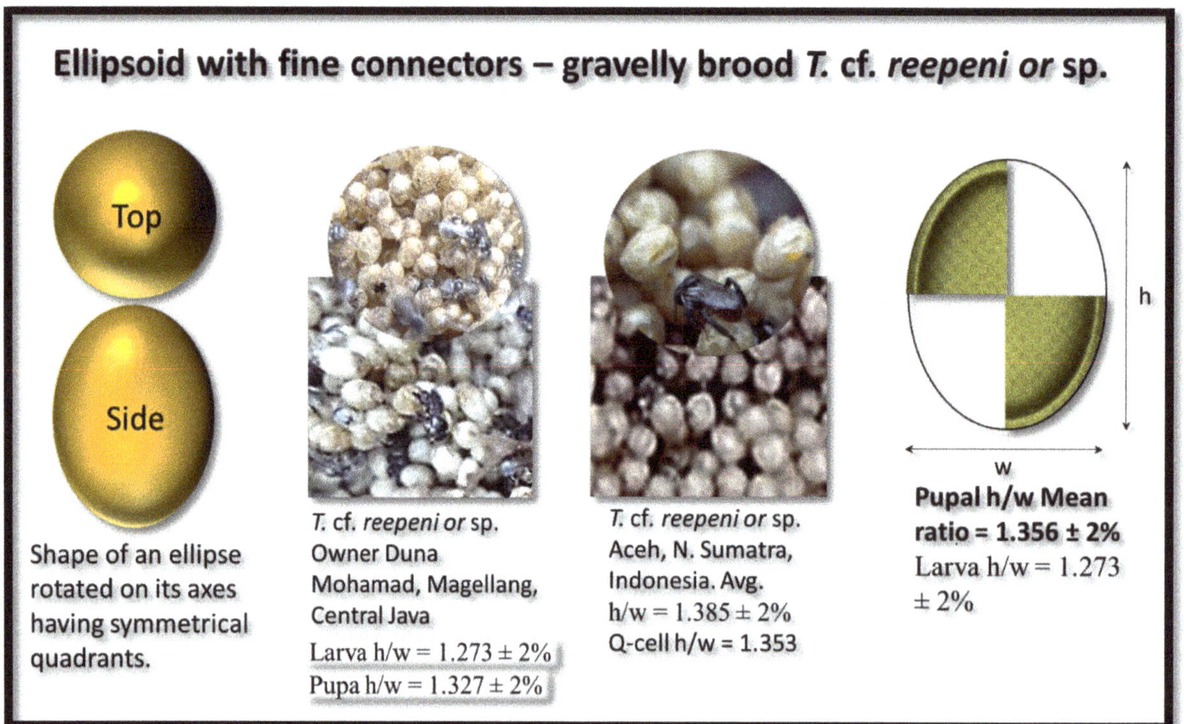

Chart 30 Ellipsoid with fine connectors – gravelly brood

14.3.2a Ovaloid – Perforated clustered brood - *Tetragonula* cf. *pagdeni* or sp.

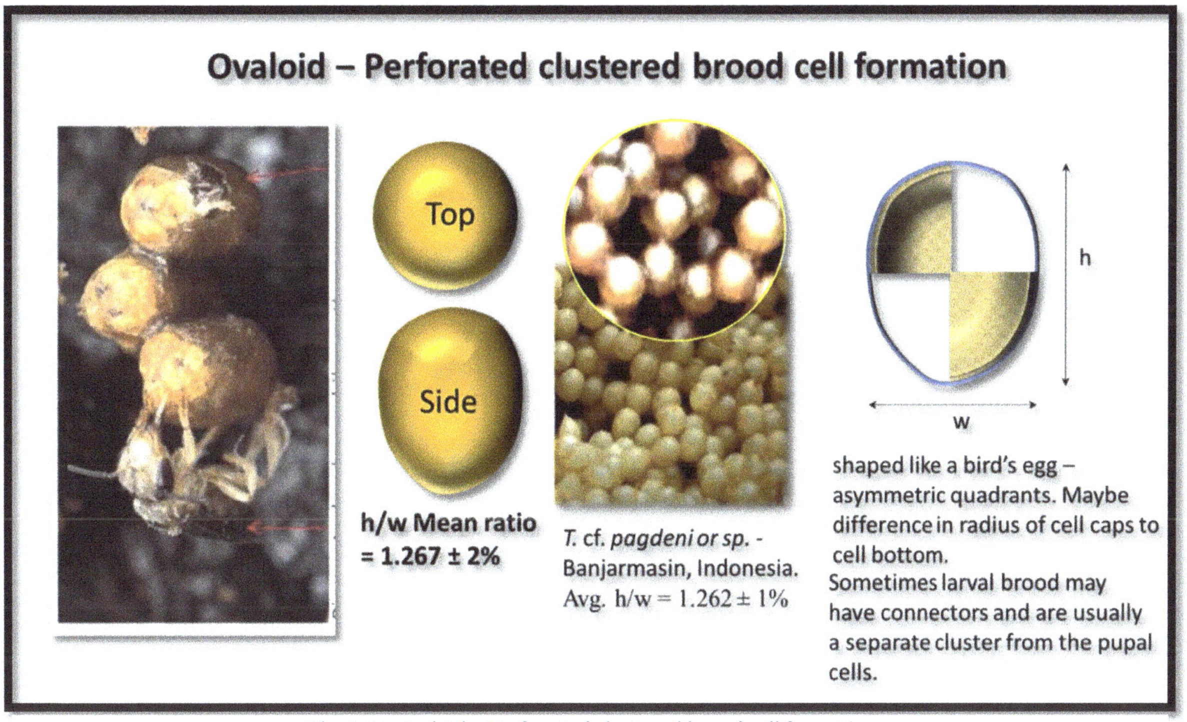

Chart 31 Ovaloid – Perforated clustered brood cell formation

14.3.2b Ovaloid -Tightly packed brood *T. melanocephala*

Chart 32 Ovaloid – Tightly packed brood T. melanocephala

14.3.3a Prolate spheroid Ω connectors[20] - ***Tetragonula cf. pagdeniformis* or *sp.***

Chart 33 Tetragonula cf. pagdeniformis or sp.

14.3.3b Prolate spheroid Ω fine connectors - ***T.* cf. *fuscobalteata* or sp.)**

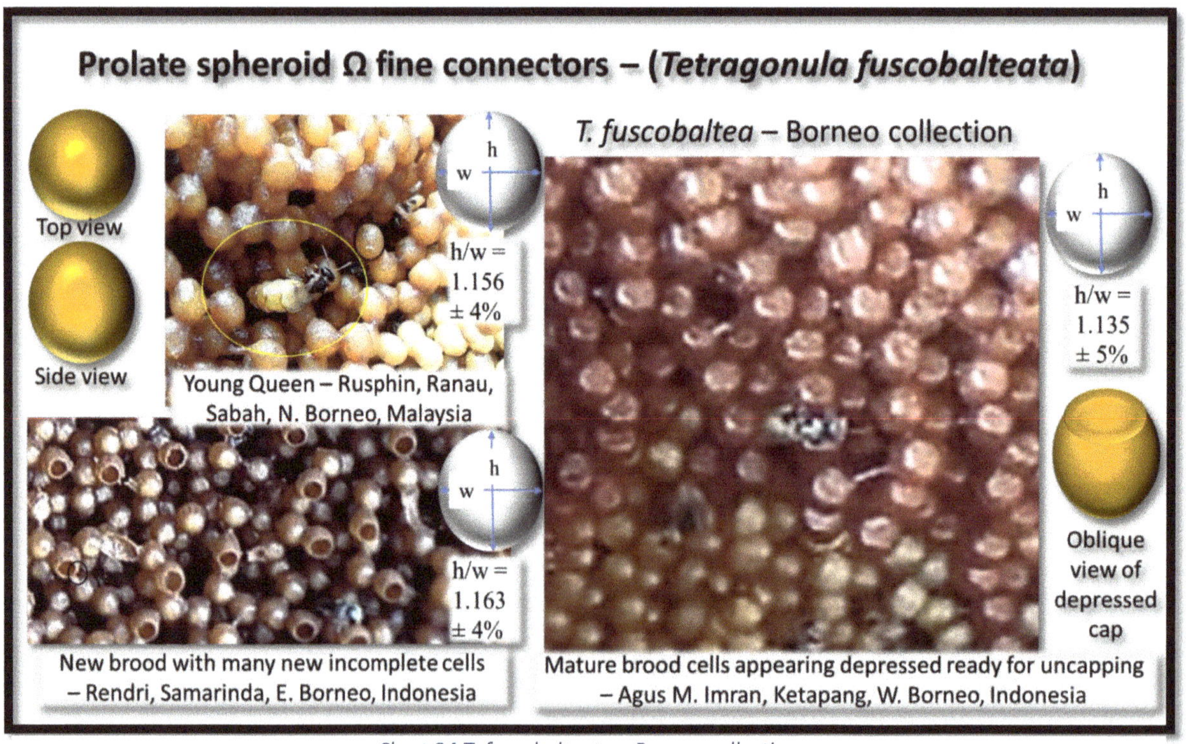

Chart 34 T. fuscobalteata – Borneo collection

---

[20]AHJ: the reddish cerumen, I believe it is from the resin of Red Balau, a *Rubroshorea* https://www.lesserknowntimberspecies.com/species/balau-red

14.3.4 Oblate spheroid clusters and semi-comb – *T. drescheri*

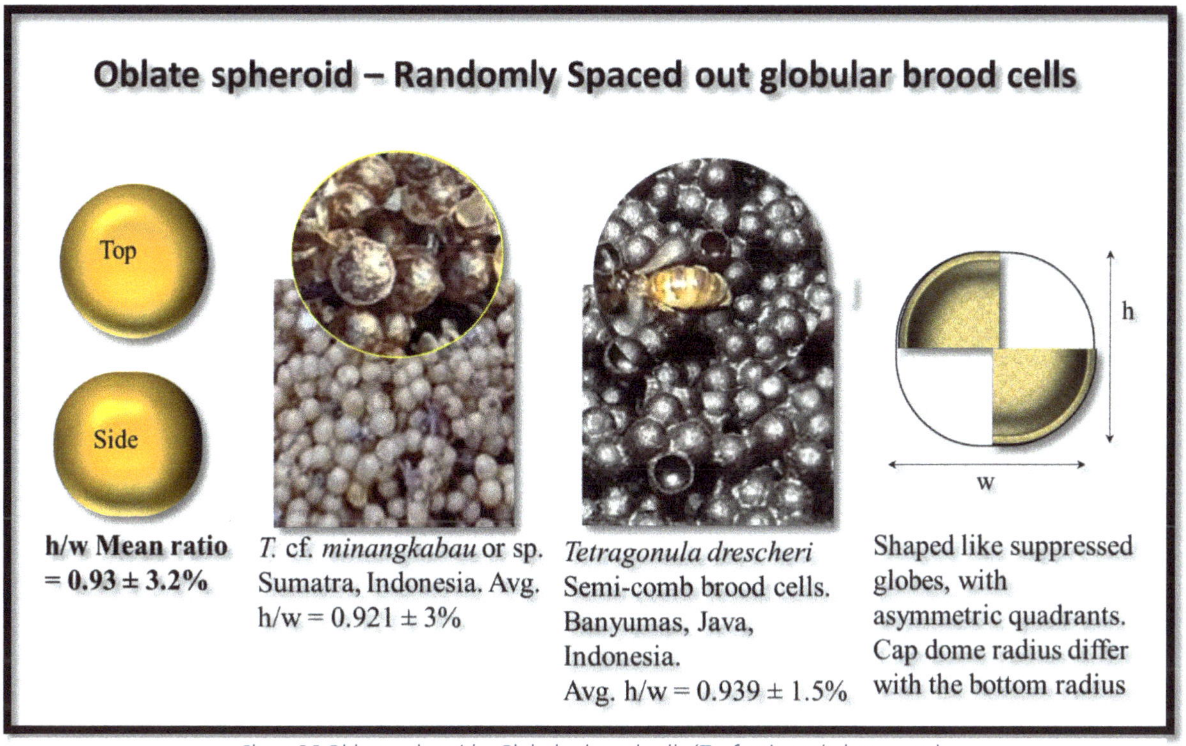

Chart 36 Oblate spheroid – Globular brood cells (T. cf. minangkabau or sp.)

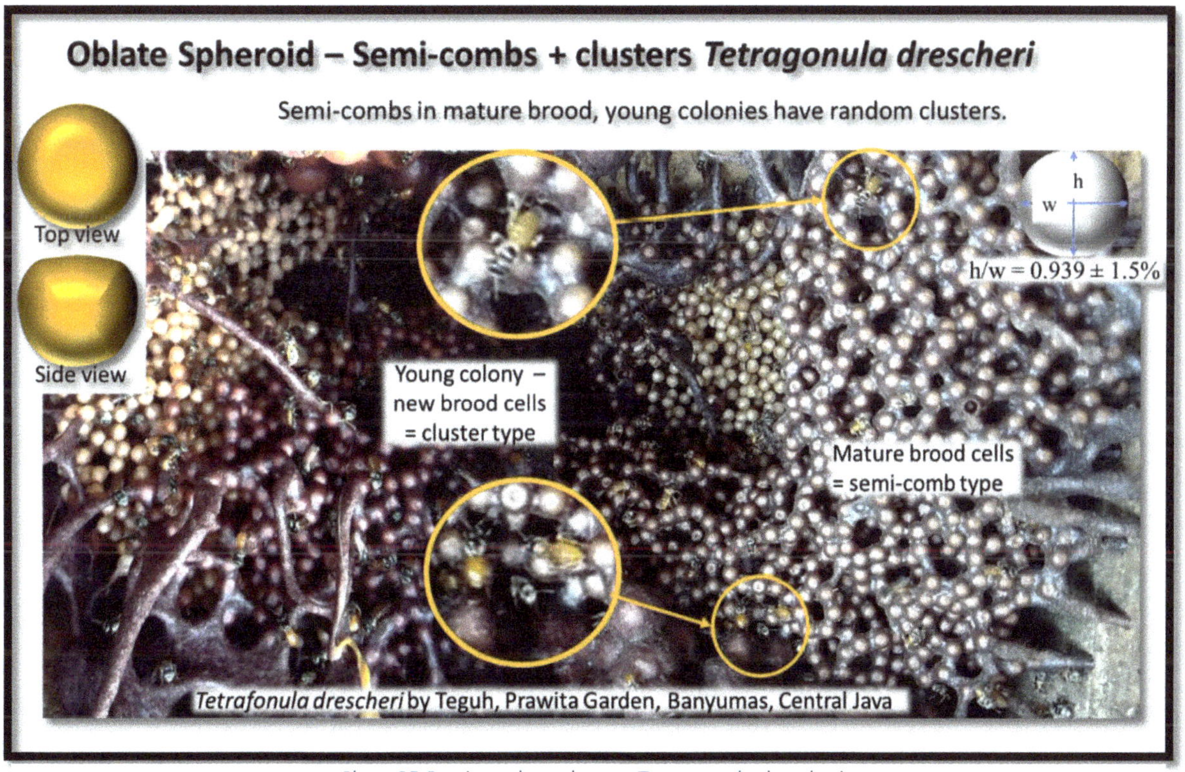

Chart 35 Semi-combs + clusters Tetragonula drescheri

14.3.5a Mixed Spheroid & Ovaloid Cells Ω Indiscriminate connectors

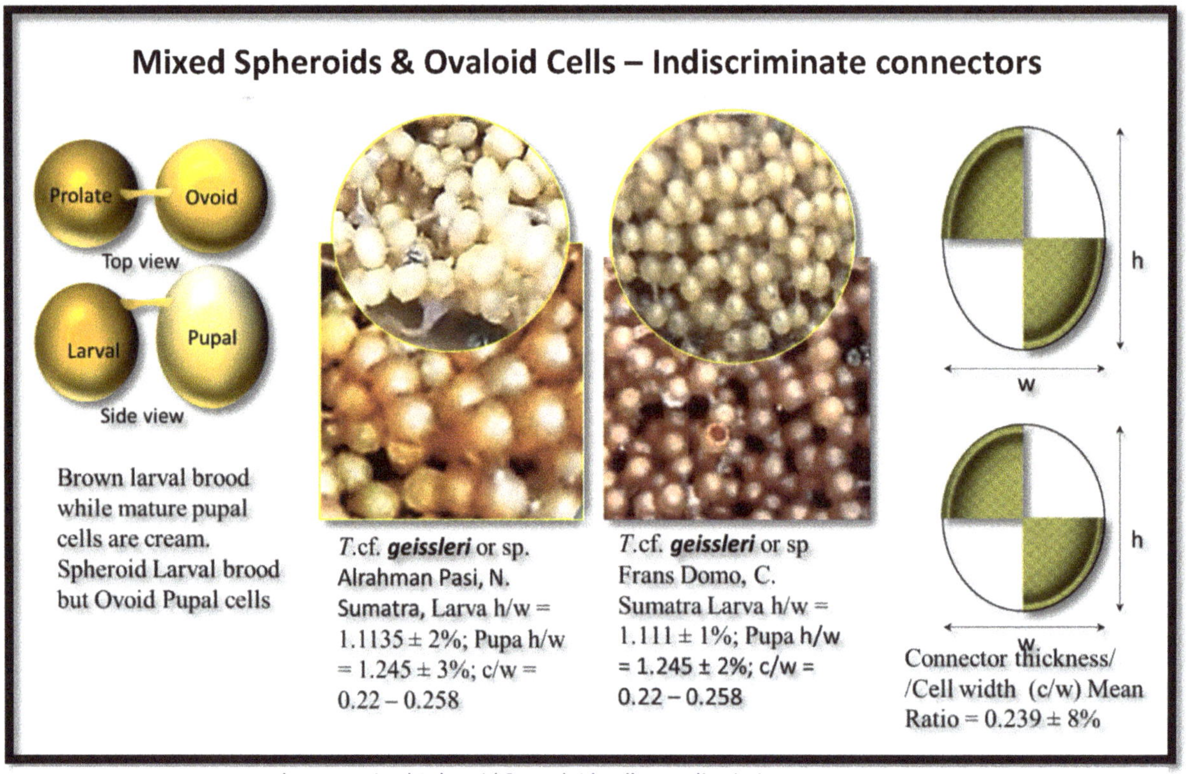

Chart 38 Mixed Spheroid & Ovaloid Cells Ω Indiscriminate connectors

Chart 37 Prolate spheroid & Ovaloid Ω fine connectors – (T. cf. geissleri or sp.)

14.3.5b Mixed Spheroid & Ovaloid Cells Ω **thick connectors**

Chart 40 Mixed Spheroid & Ovaloid Cells Ω Thick connectors

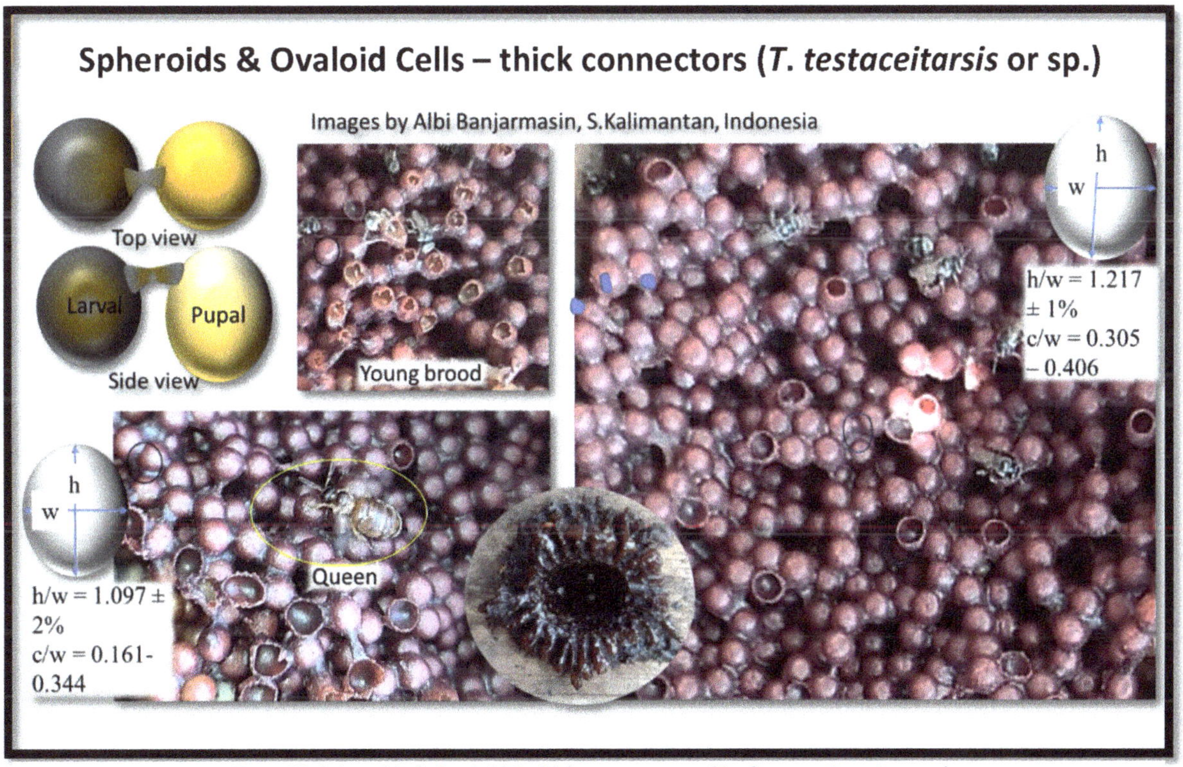

Chart 39 Spheroids & Ovaloid Ω thick connectors – (T. cf. testaceitarsis or sp.)

14.3.6a Combo Mixed Cell shapes – Ellipsoid Pupal, Ovaloid Pre-pupal & Spheroid Larval.

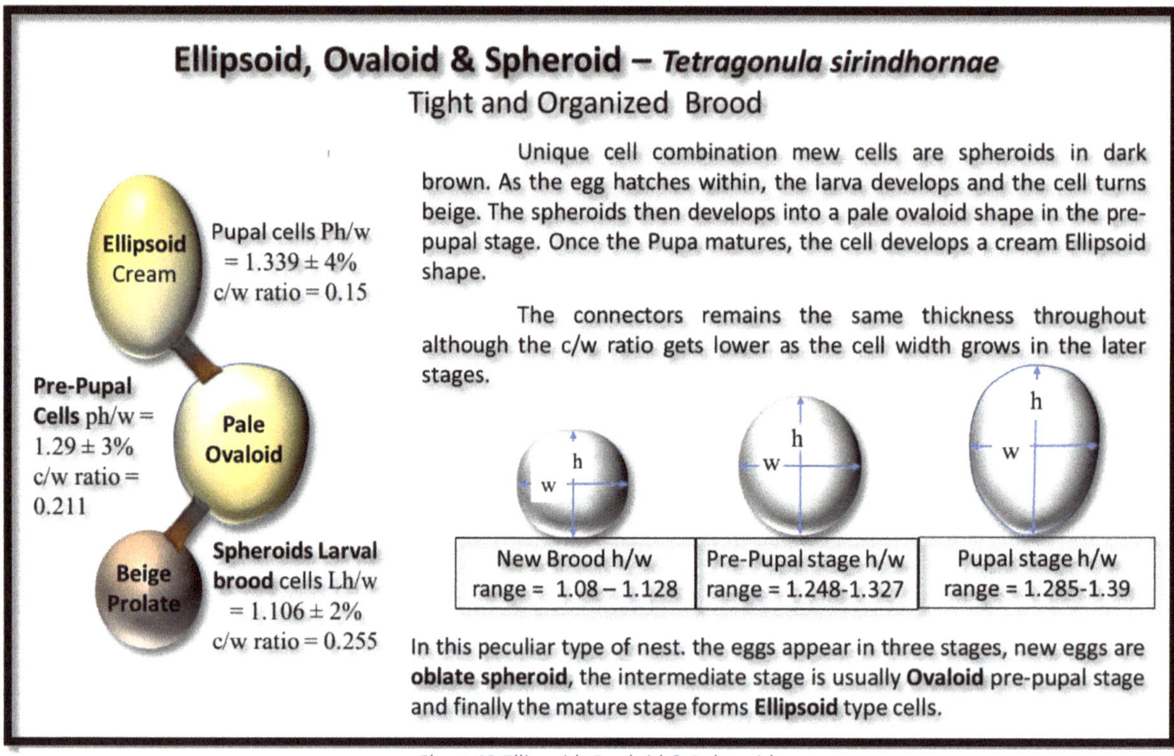

Chart 42 Ellipsoid, Ovaloid & Spheroid.

Chart 41 Tetragonula sirindhornae - Tight and Organized Brood.

14.3.6b Combo Mixed Cell shapes - Ellipsoid, Ovaloid and Spheroid – loose brood structured - *Tetragonula (Tetragonilla) collina.*

Chart 44 Ellipsoid, Ovaloid and Spheroid – loose brood structured

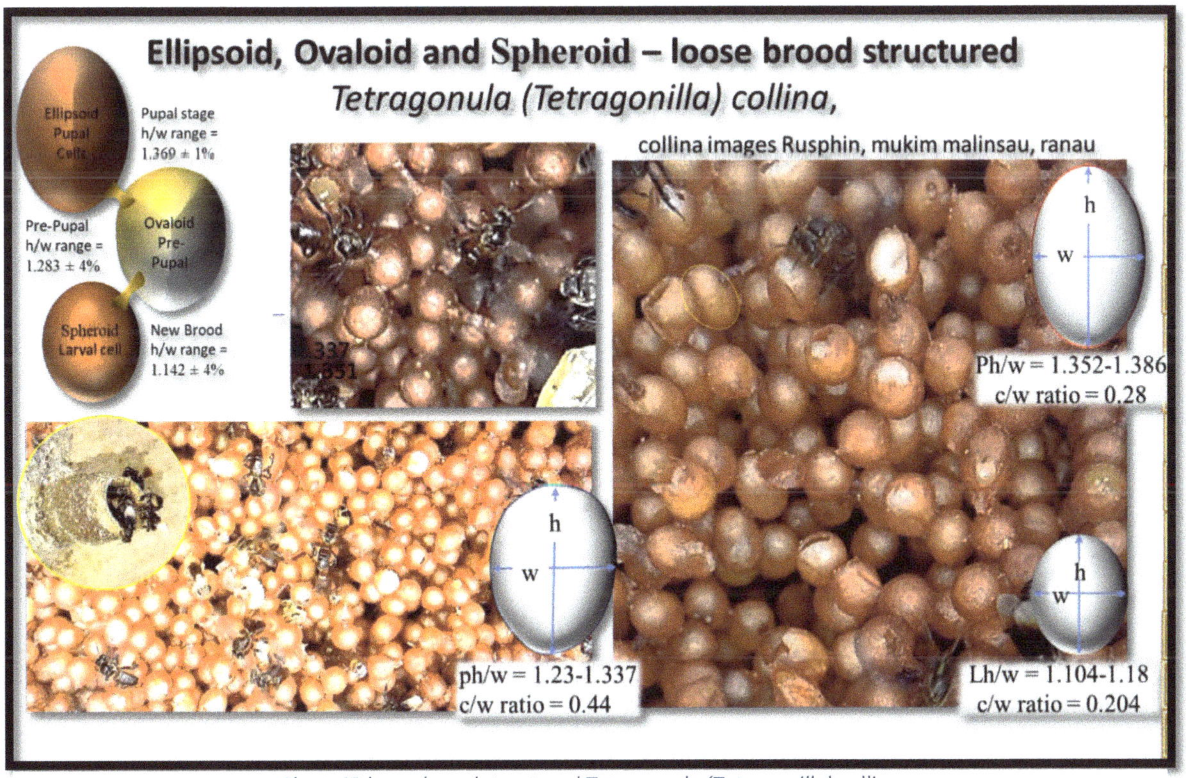

Chart 43 loose brood structured Tetragonula (Tetragonilla) collina,

14.3.6c Combo Mixed Cell shapes - Spheroid Larval & Pre-Pupal + Cylindroid Pupal

Chart 45 Larval & Pre-Pupal + Cylindroid Pupal Tetragonula sp. Palawan, Philippines

Section 4: Wallacea - Philippines Oceania & Indonesian Wallacea

14.4.1 Ellipsoidal - Compact or tight brood – *T. laeviceps*

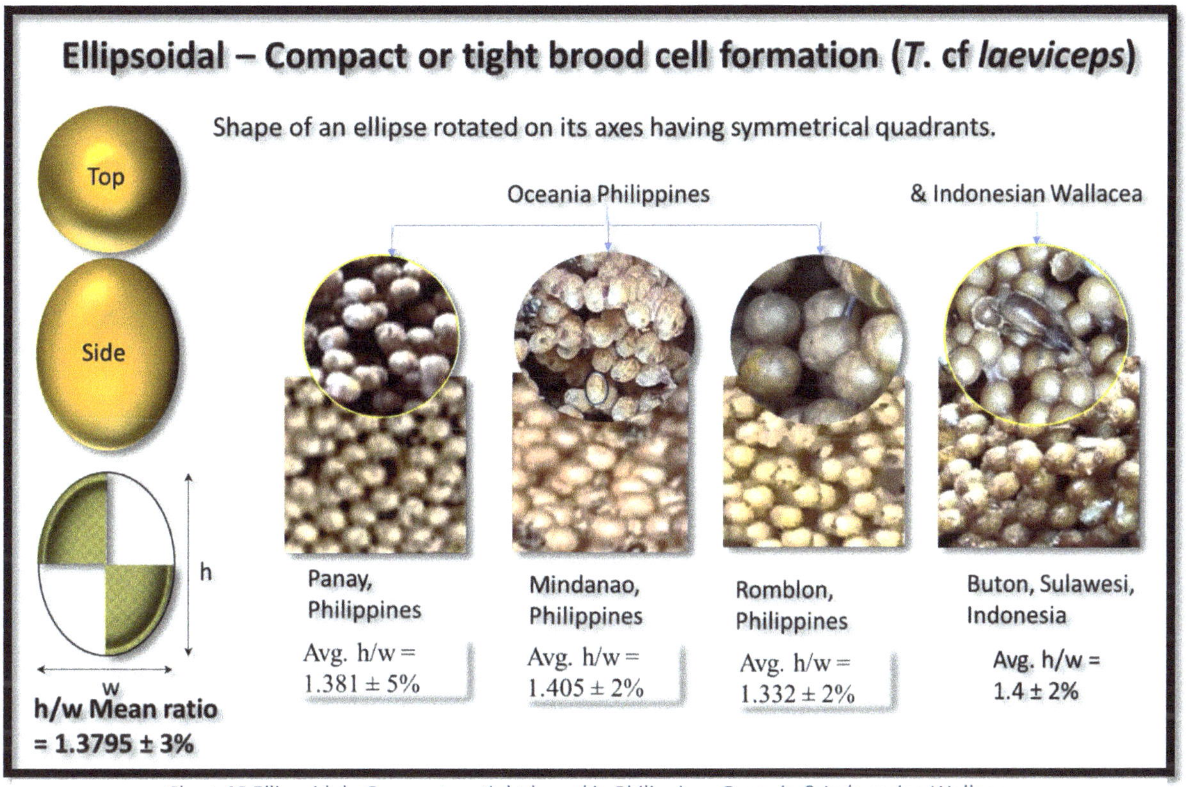

Chart 46 Ellipsoidal - Compact or tight brood in Philippines Oceania & Indonesian Wallacea

14.4.2a Ovaloid - Sparse clumps formation - *T.* cf. *iridipennis* or sp.

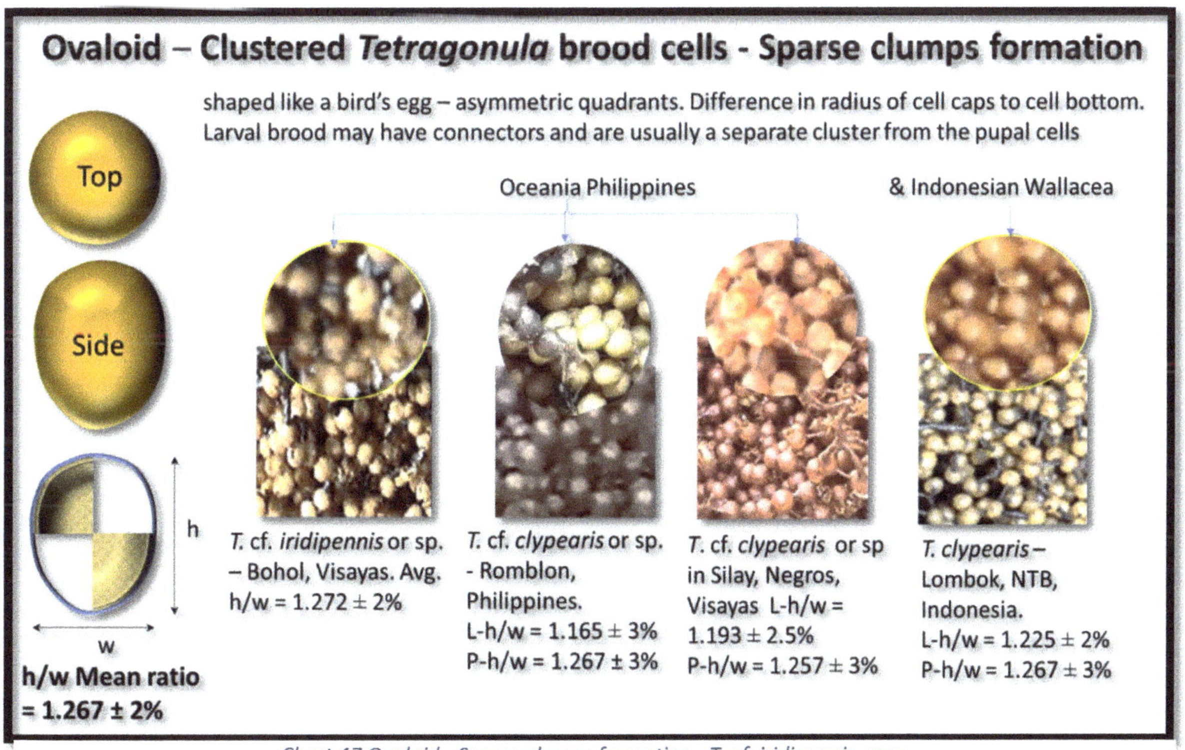

Chart 47 Ovaloid - Sparse clumps formation - T. cf. iridipennis or sp.

14.4.2b Ovaloid - Sparse clumps formation - *T.* cf. *clypearis* or sp

Chart 48 T. cf. clypearis or sp in Silay, Negros, Visayas

14.4.3a Prolate spheroid Ω fine connectors - *T.* cf. *dapitanensis* or sp

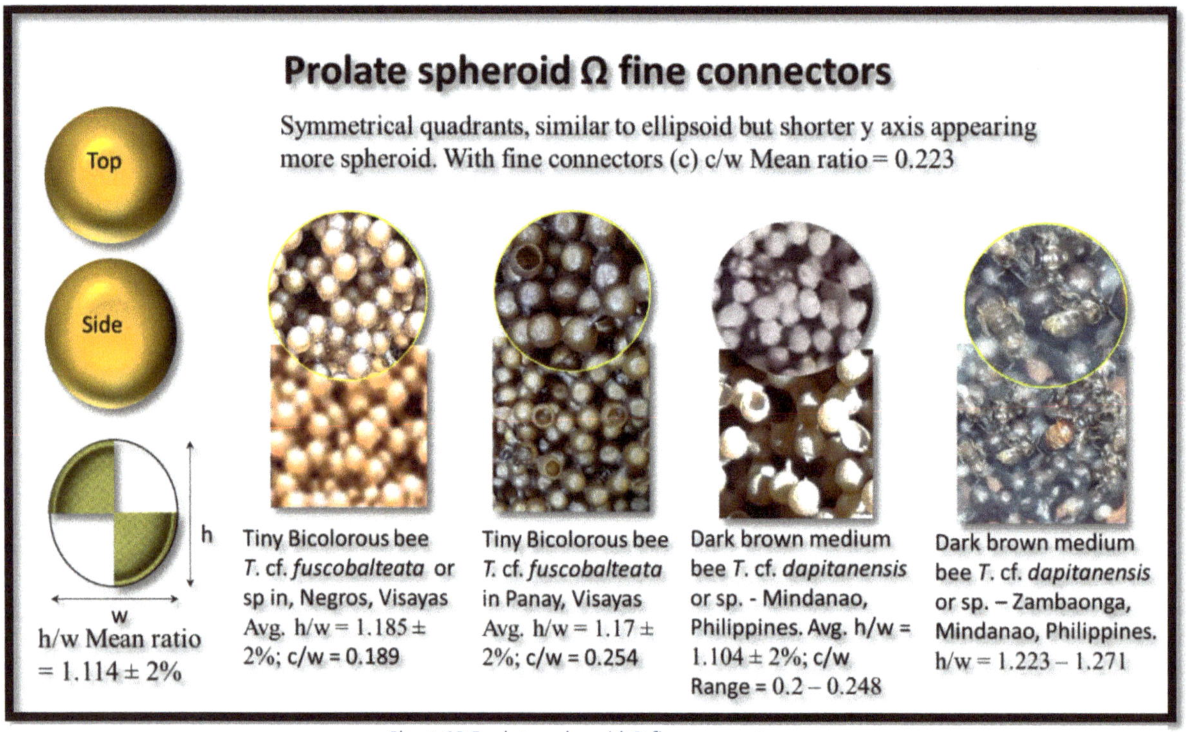

Chart 49 Prolate spheroid Ω fine connectors

14.4.3b Prolate spheroid Ω fine connectors - *T. cf. fuscobalteata*

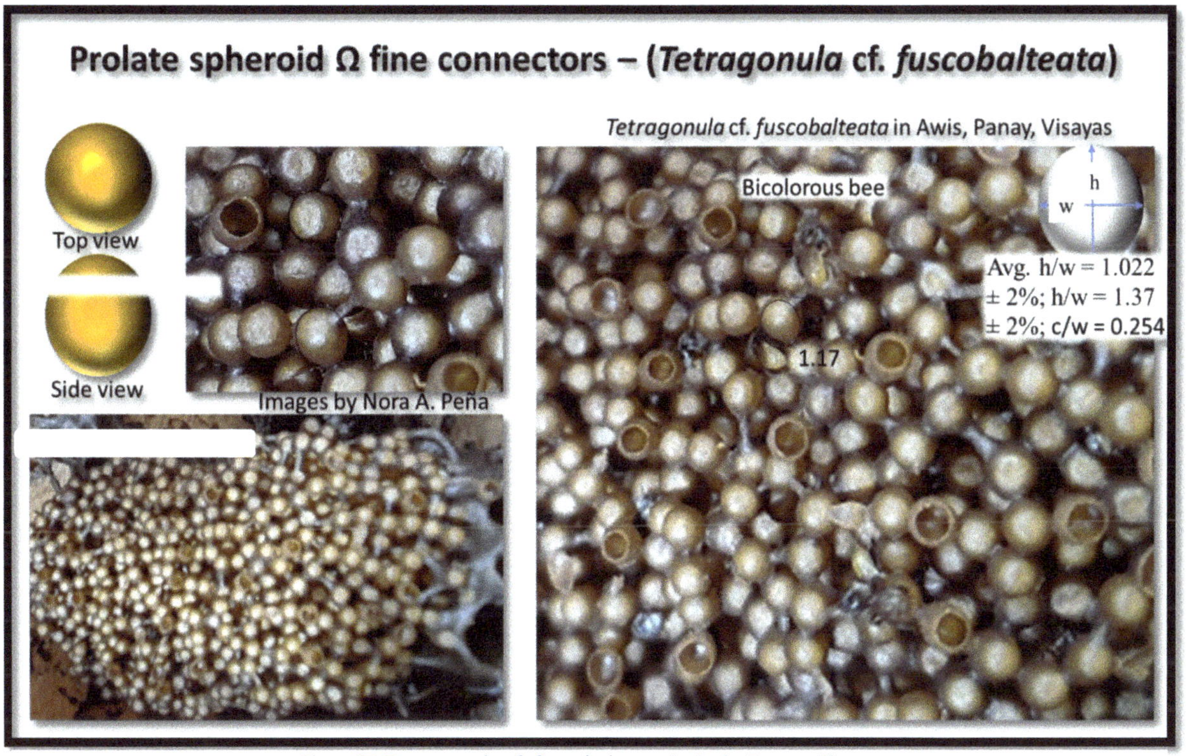

Chart 50 Tetragonula cf. fuscobalteata in Awis, Panay, Visayas

14.4.4 Combo Pairs Spheroid, Ovaloid or Cylindroid Cells Ω connectors

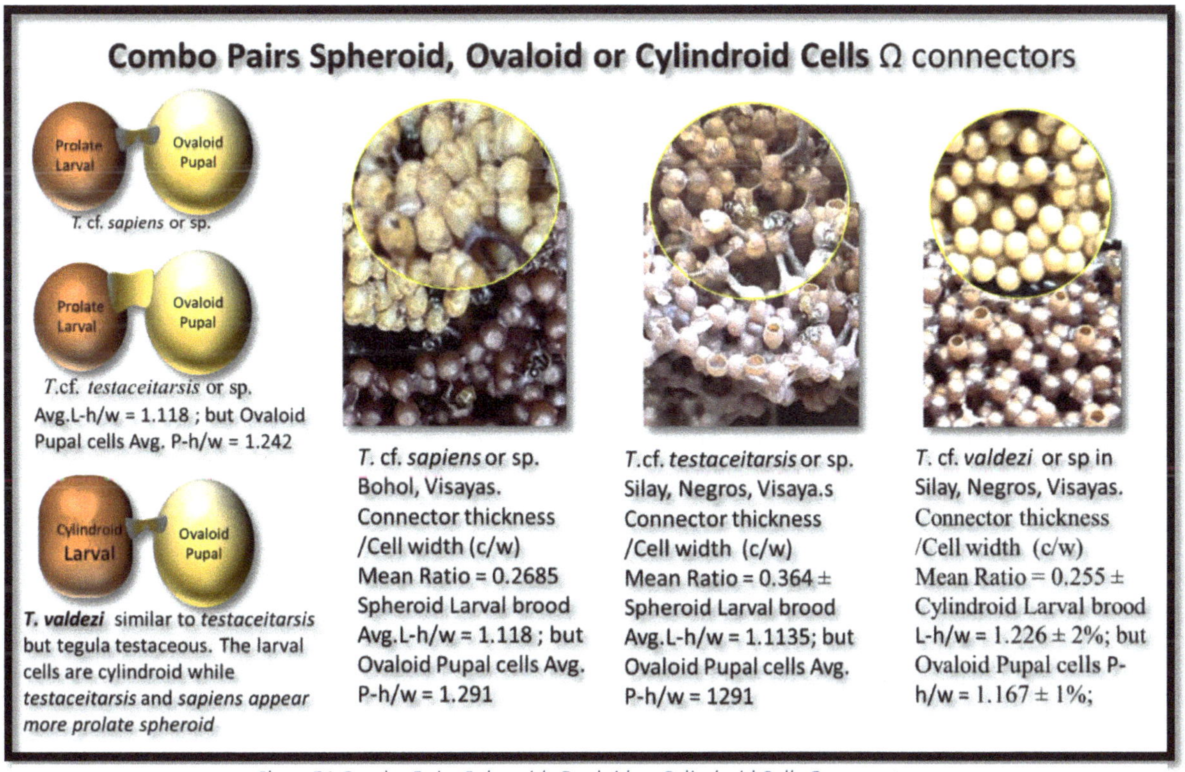

Chart 51 Combo Pairs Spheroid, Ovaloid or Cylindroid Cells Ω connectors

14.4.5 Combo Triads Oblate Spheroid, Prolate Spheroid & Ovaloid Cells

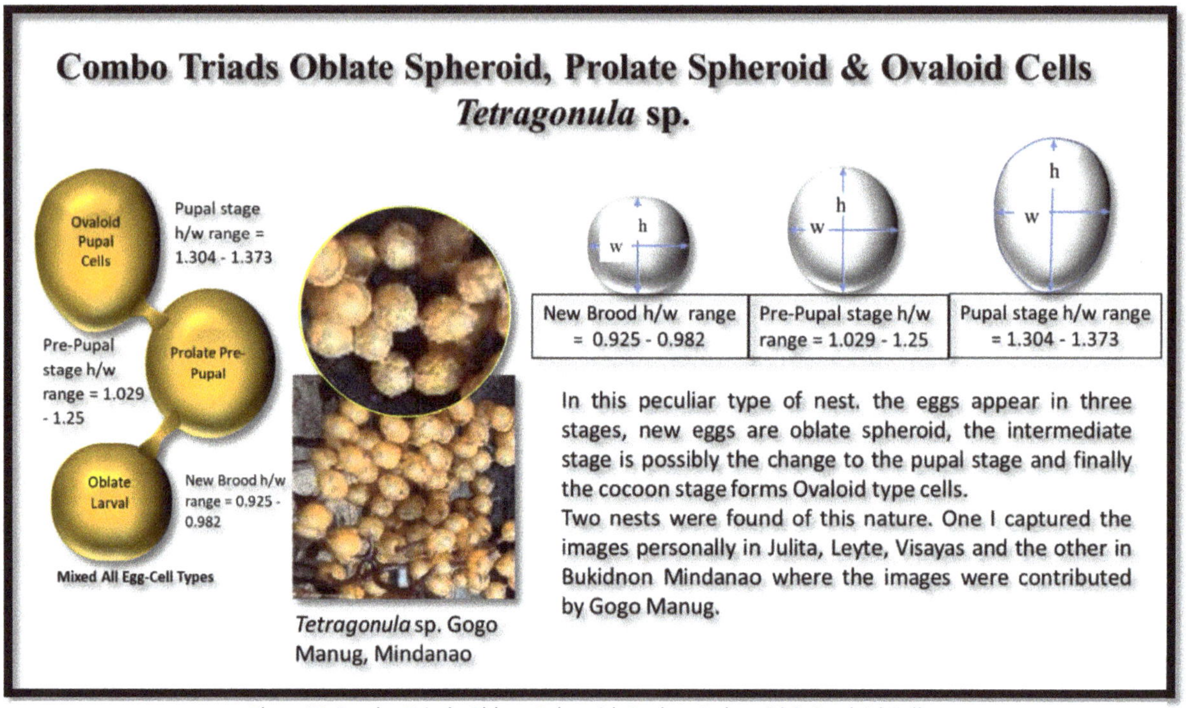

Chart 53 Combo Triads Oblate Spheroid, Prolate Spheroid & Ovaloid Cells

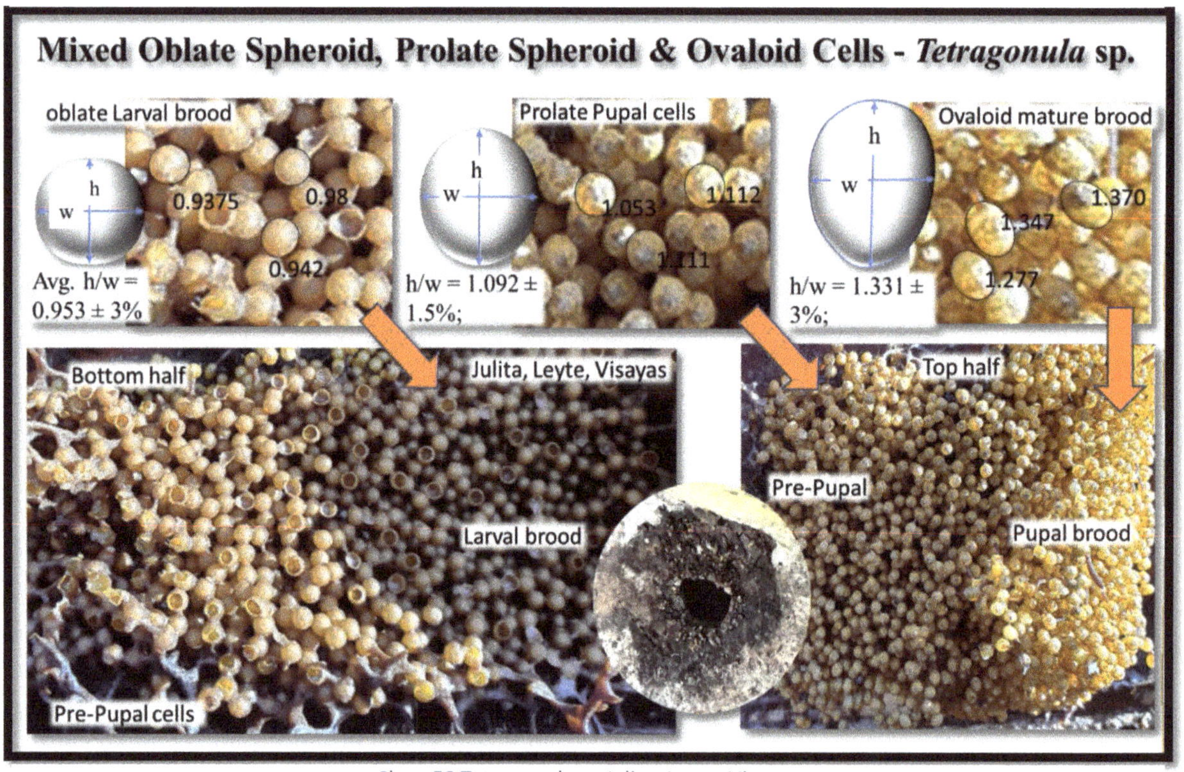

Chart 52 Tetragonula sp Julita, Leyte, Visayas

14.4.6 Cylindroid layered comb brood cells

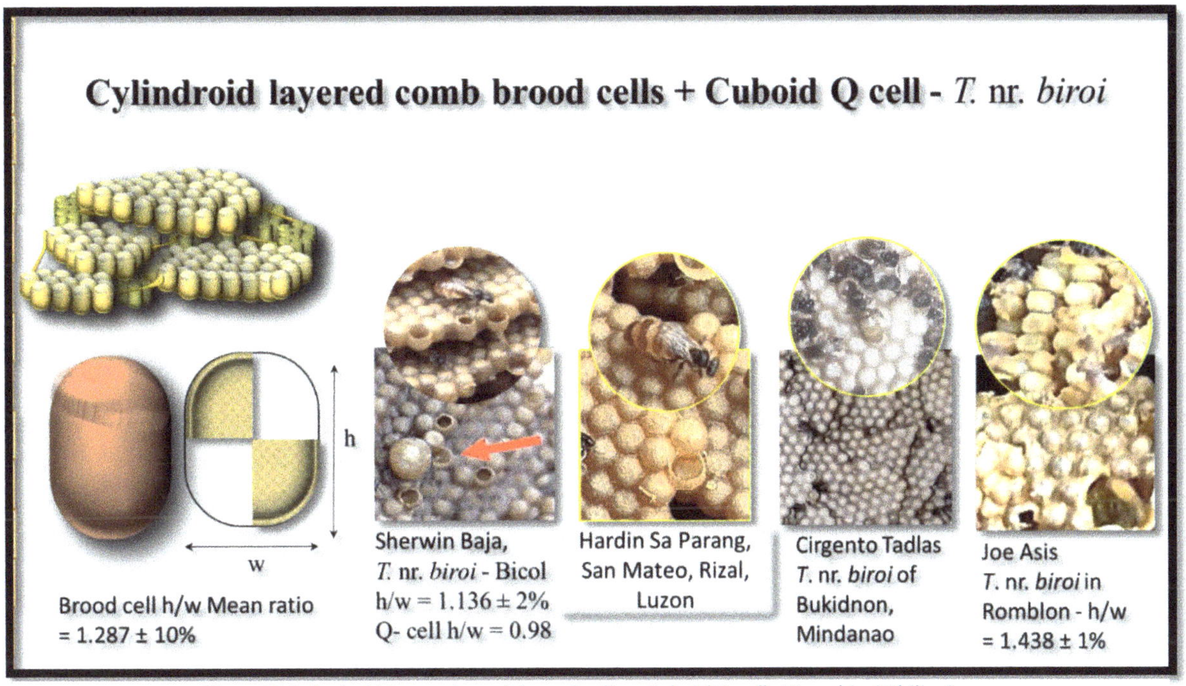

Chart 55 Cylindroid layered comb brood cells + Cuboid Q cell - T. nr. biroi Philippines

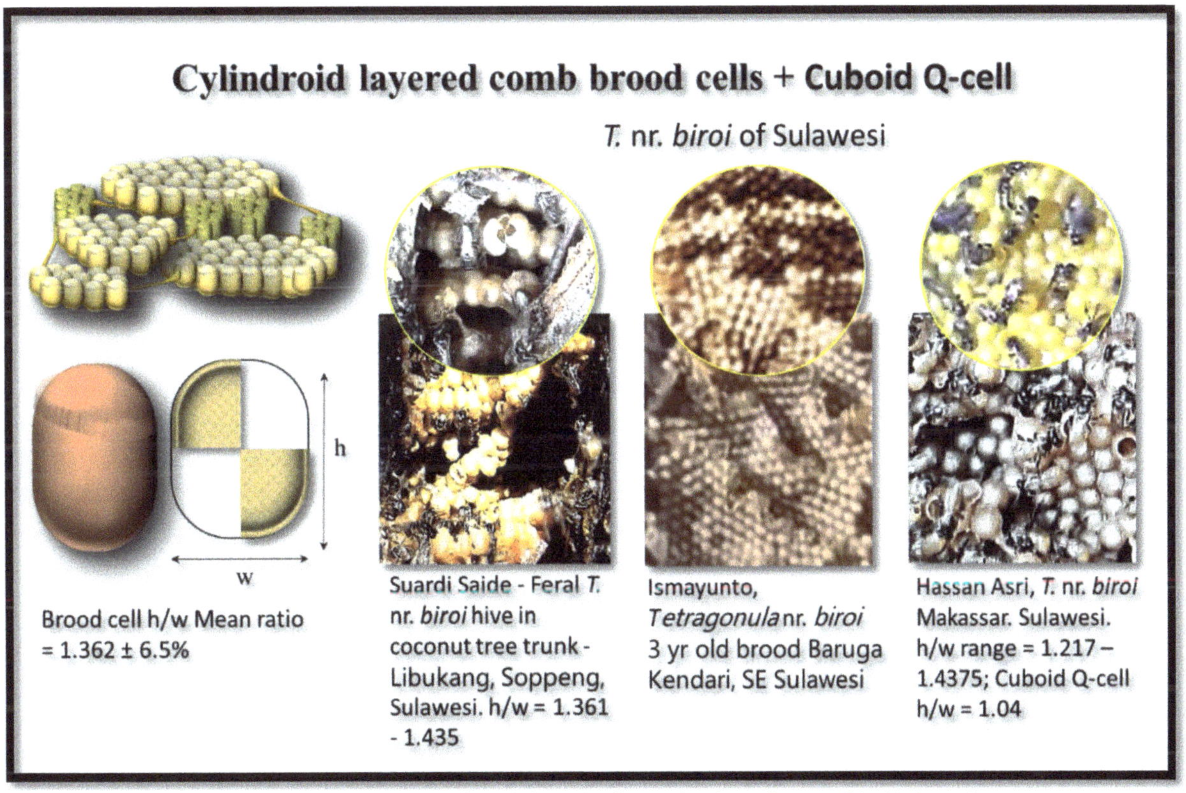

Chart 54 Cylindroid layered comb brood cells + Cuboid Q cell - T. nr. biroi Sulawesi

Wallacetrigona Brood cells in **Philippines Oceania & Indonesian Wallacea**

14.4.7 Cylindroid layered comb brood cells - *Wallacetrigona incisa*

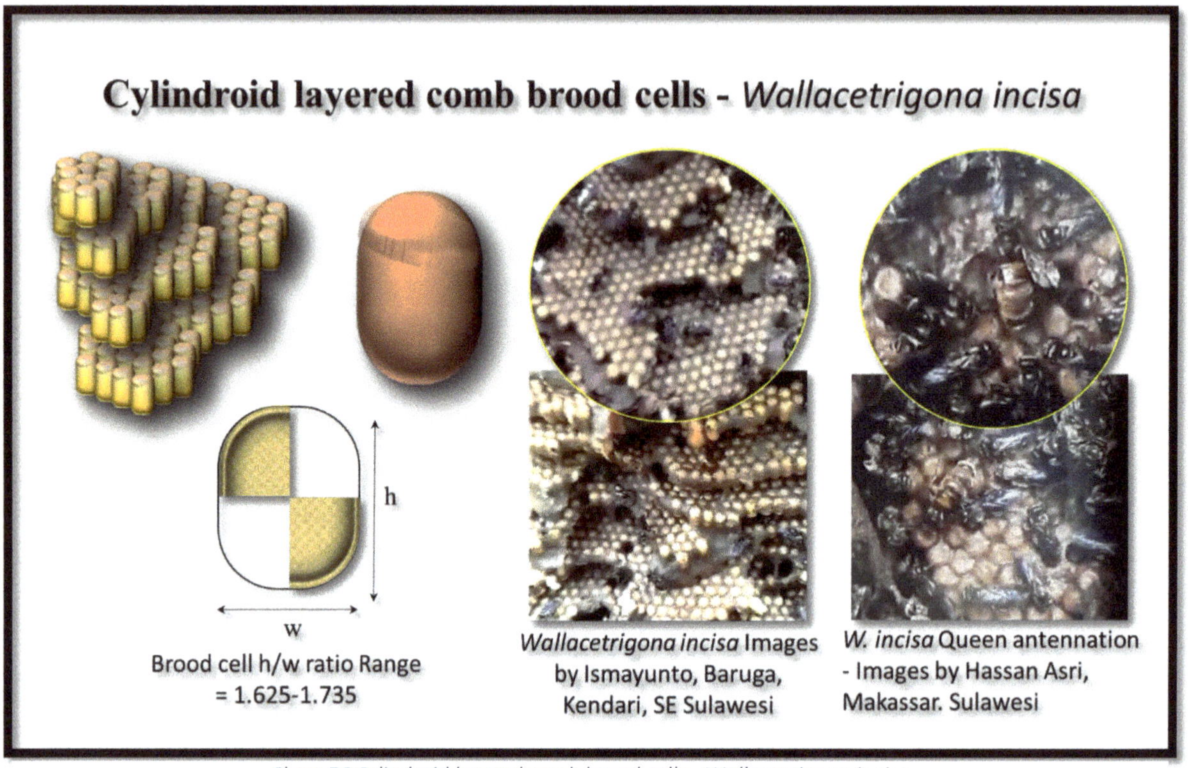

Chart 56 Cylindroid layered comb brood cells - Wallacetrigona incisa

Section 5: Sahul Shelf

14.5.1 Ellipsoidal - Compact or tight brood clusters

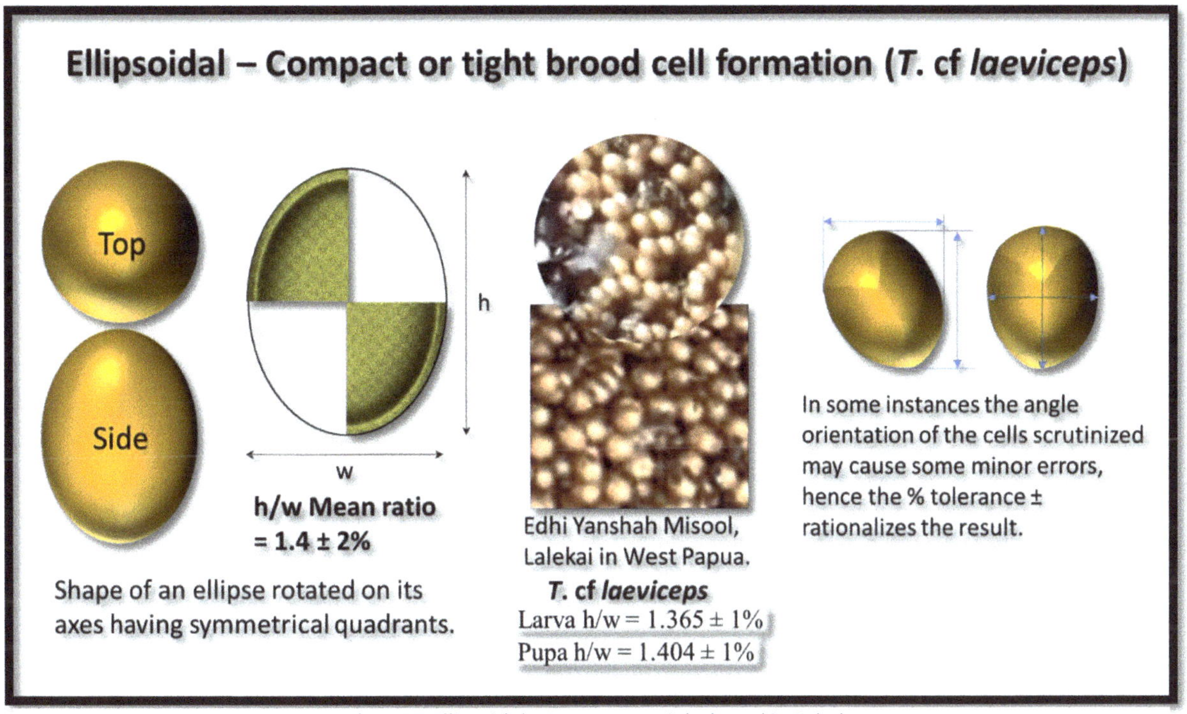

Chart 57 Ellipsoidal - Compact or tight brood in Sahul

14.5.2 Ovaloid – Sparse Clustered brood cell clumps

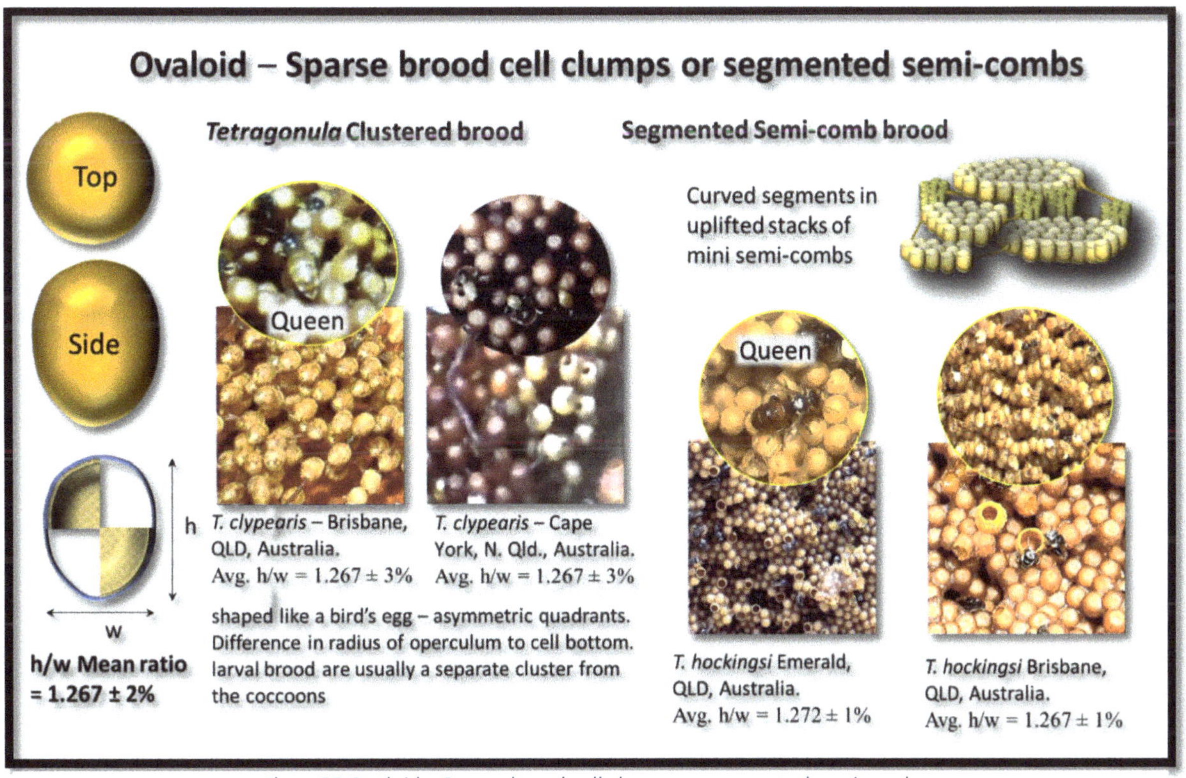

Chart 58 Ovaloid – Sparse brood cell clumps or segmented semi-combs

14.5.3 **Ovaloid & Prolate spheroid** – Graduated brood

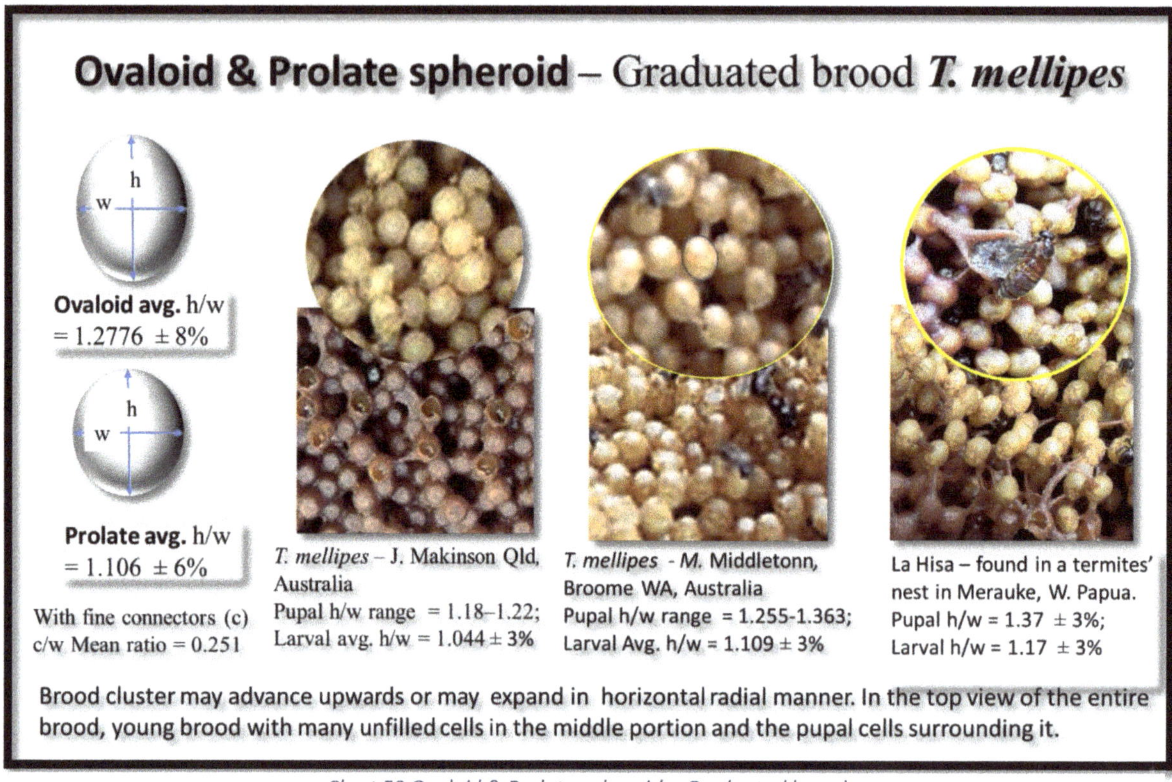

Chart 59 Ovaloid & Prolate spheroid – Graduated brood

14.5.4 Cylindroid – semi-comb brood with or without filaments

Chart 60 Cylindroid – semi-comb brood with or without filaments

14.5.5a Mixed Oblate & Prolate Spheroid Cells – Papua New Guinea

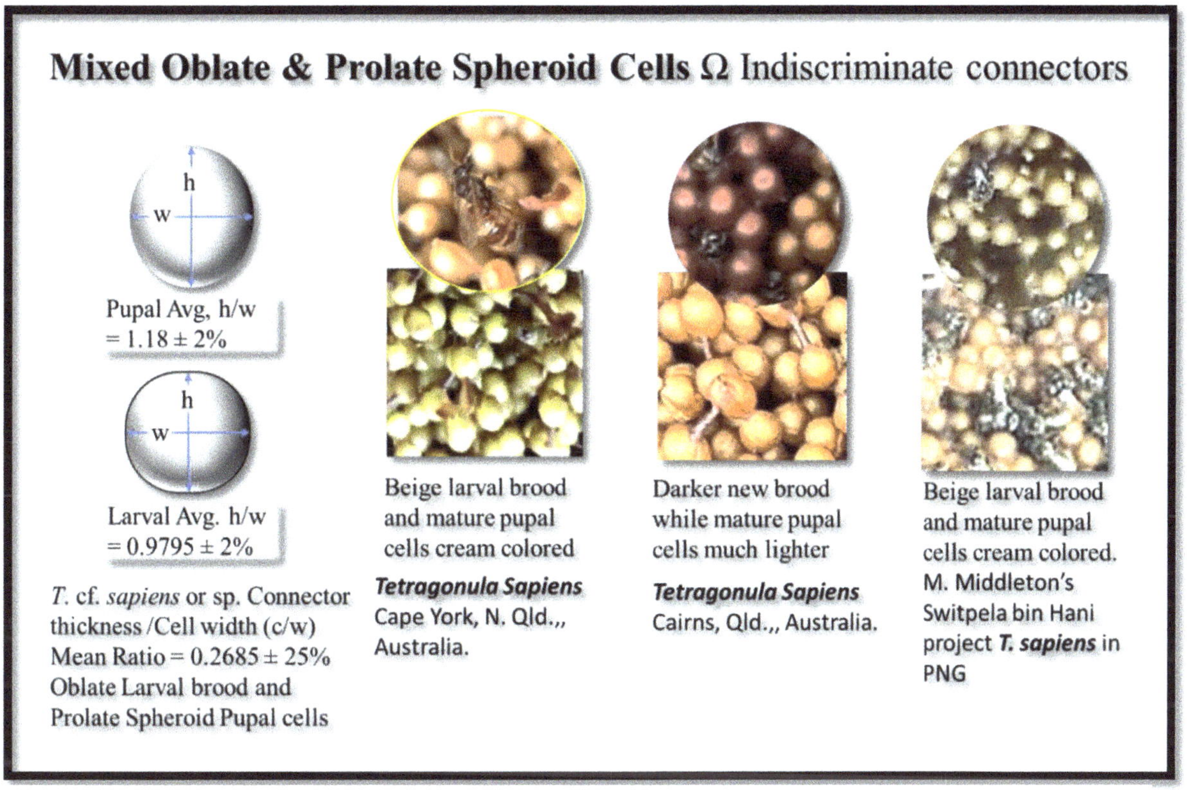

Chart 62 Mixed Oblate & Prolate Spheroid Cells Ω Indiscriminate connectors

14.5.5b Mixed Oblate & Prolate Spheroid Cells – Australia

Chart 61 Oblate & Prolate Spheroids Ω connectors - Tetragonula sapiens – N. Qld., Australia

14.5.6 Cylindroid – Spiralled or layered stack brood cell combs

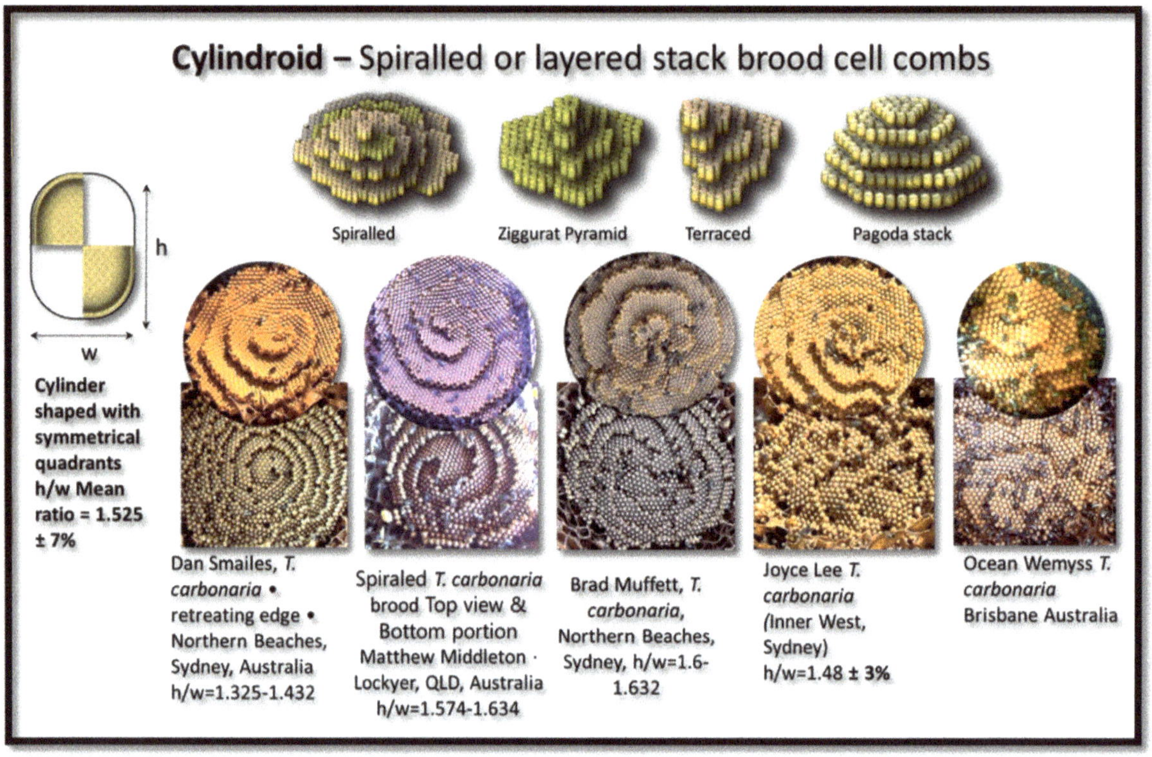

Chart 65 Cylindroid – Spiralled or layered stack brood cell combs

Chart 64 Brood formation structural sequence - Tetragonula carbonaria

Section 6: Other Genera in the Sahul Shelf

14.6.1 Prolate spheroids – *(Austroplebeia australis)*

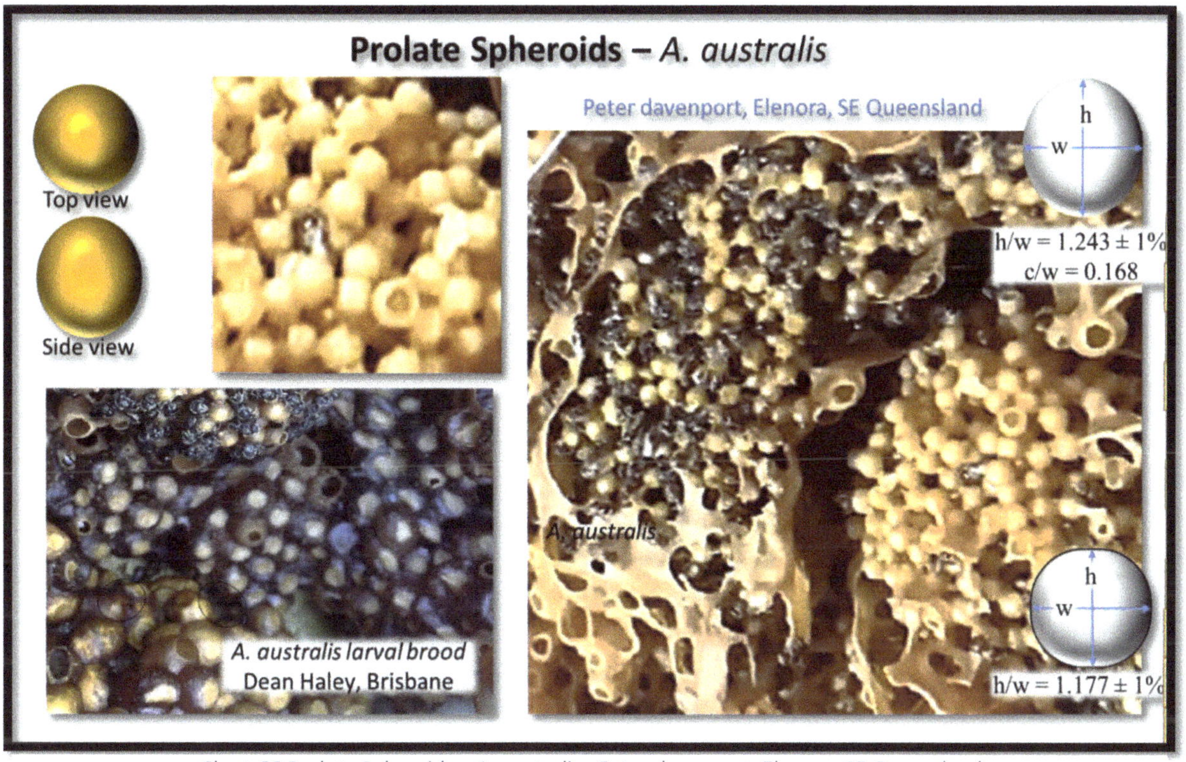

Chart 66 Prolate Spheroids – A. australis - Peter davenport, Elenora, SE Queensland

14.6.2 Prolate spheroids - *(Austroplebeia cassiae)*

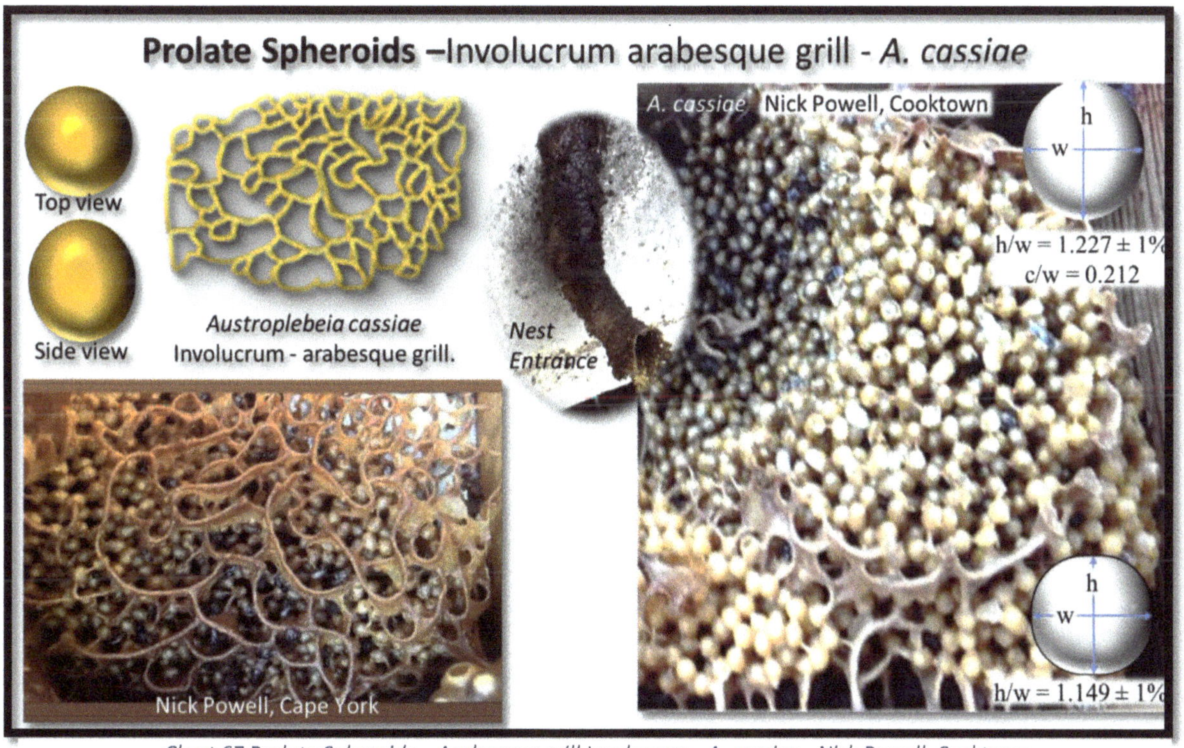

Chart 67 Prolate Spheroids – Arabesque grill Involucrum - A. cassiae - Nick Powell, Cooktown

14.6.3a Prolate spheroids - (*Austroplebeia. cincta)* - Australia

Chart 68 Prolate Spheroids – Fluffy Involucrum – A. cincta - Dan-lyn Evans, Weipa, Qld

14.6.3b Prolate spheroids - (*Austroplebeia. cincta)*- Papua

Chart 69 Prolate Spheroids – Variable Involucrum – A. cincta – La Hisa, Merauke, West Papua

Heterotrigona (Platytrigona) Brood cells in the Sahul Shelf

14.6.4 Ovaloid clusters - *Heterotrigona (Platytrigona) cf. flaviventris*, Papuasia

Chart 70110 Ovaloid – Sparse random brood cell clusters Heterotrigona (Platytrigona) cf. flaviventris, Papua

Epilogue

As it turns out, this book has more images than a coffee table book. The idea was that a picture paints a thousand words to avoid misconceptions and misunderstandings from the written word. The bees did not leave behind literature to comprehend their behaviour and architecture, so we had to piece together our observations of their daily lives. It covers the history of their (*Tetragonula*) nascency and present-day nest architecture in their domesticated boxes or numerous other shaped hive containers.

These observations are not easily described seeing that there are hundreds of languages and dialects in the ASEAN region. It would take ages for one observation alone to be described and then translated into the languages in the relevant regions. As it is aimed at beekeepers primarily and in this modern internet age, images of such observations tell a more resounding story and meaning. The images were collected and contributed by many beekeepers in the social media of similar beekeeping interest groups. These were necessary due to movement restrictions, as this book was written during the Covid19 pandemic.

We did manage to classify the brood cell shapes and brood form after sorting out hundreds of images from the Indo-Malayan and Australasian ecozones. Emphasis was given to *Tetragonula* as the most widespread genus of stingless bees in the ASEAN region. This brood cell shape classification and brood form categorization will help many beekeepers and enthusiasts recognize the bees they keep and subsequently collect and harvest their products by the classifications and categories established.

Students and researchers may expand and be guided by these preliminary findings and determine the extent of the validity. The results may then form a supplementary criterion for identifying the species. Not only that, but also whether the climate conditions or changes affect the brood forms. For example, the situation where we find four different forms of brood architecture in a *T.* nr. *biroi* colony in a box. The question arises whether the phenomenon occurs due to domestication efforts or whether climate change. Another example is the regimented brood (page 75) that happens only on the box's sidewall, and we do not see regimented brood in feral hives.

Other peculiarities worth mentioning here are the many varied brood forms seen in a domesticated hive, especially in Vietnam. Are they complex species, or are they different species altogether? Much DNA analysis is required and would answer many puzzles encountered.

Lastly, I end with an illustrated summary of brood forms and an algorithm indicating the key to Indo-Malayan and Australasian Genera and the Subgenera of Meliponini based on Rasmussen, Thomas & Engel, 2017.

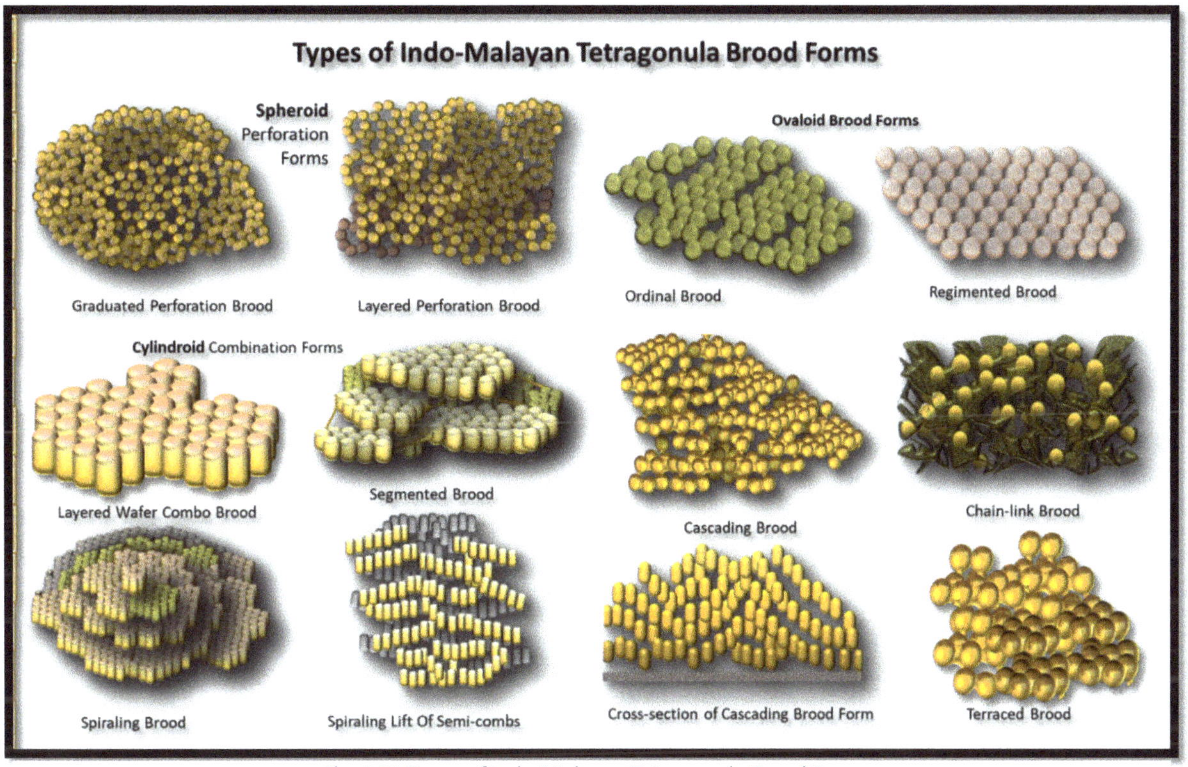

Chart 71 Types of Indo-Malayan Tetragonula Brood Forms

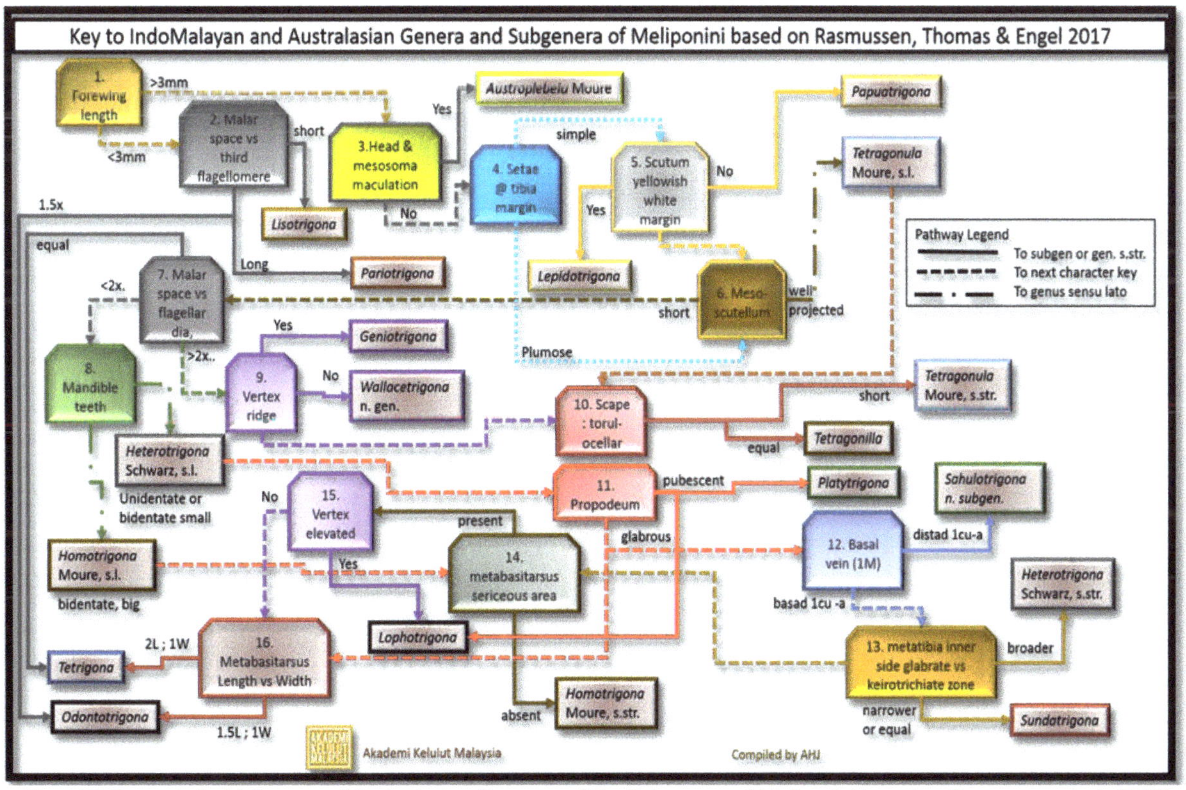

Chart 72 Key to Indo-Malayan and Australasian Genera and the Subgenera of Meliponini based on Rasmussen et al., 2017.

List of Figures, Portraits, Charts & Tables

List of Charts

List of Figures

List of Portraits

List of Tables

List of Maps

Bibliography

Baja, S. M., & Mostoles, M. D. (2017). Colony Characterization And Amplicon Sequencing Of Stingless Bees (*Tetragonula biroi*). In A. H. Jalil, *Handbook of Meliponiculture vol 2.* Akademi Kelulut Malaysia.

Baltazar, C. R. (1966). A Catalogue Of Philippine Hymenoptera. *Pacific Insects Monograph 8.*

Boongird, S. &. Michener, C. D.. (2004). A New Species of *Trigona* from Peninsular Thailand (Hymenoptera: Apidae: Meliponini).April 2004 *JKES* 77(2):143-146

Boontop, Y., Malaipan, S., & Wiwatwittaya, K. C. (2008). Diversity of Stingless Bees (Apidae: Meliponini) in Thong Pha Phum District, Kanchanaburi Province, Thailand. *Kasetsart J. (Nat. Sci.) 42*, 444 - 456.

Chandra, G. (n.d.). *Wallace's Line, Weber's Line and Wallacea.* Retrieved from IASZoology.com: http://www.iaszoology.com/wallacea/

Chinajariyawong, A., & Saiboon, S. (n.d.). Culture of the stingless bee, Trigona laeviceps Smith (Hymenoptera: Apidae). *Institute of Agricultural Technology, Walailak University.*

Chinh, T., Sommeijer, M., Boot, W., & Michener, C. (2005). Nest Architecture and Colony Characteristic of Three Stingless bees in North Vietnam with the First Description of the Nest of *Lisotrigona carpenter* Engel (Hymenoptera: Apidae, Meliponini). *J. Kans Entomol. Soc. 3*, 26-39.

Chuttong, B., & Burgett, M. (2017). Biometric studies of the stingless bee *Tetragonula laeviceps* species complex (Apidae: Meliponini) in SE Asia (Thailand). *1st International Meliponine Conference 2017.*

Cockerell, T. D. (1919). The Social Bees of The Philippines Islands. *Univeraity of Colarado.*

Cortopassi-Laurino, M., & al, e. (2006). Global meliponiculture: challenges and opportunities. *Apidologie 37*, 275–292.

Danaraddi, C. S., Viraktamath, S., & Bhat, K. B. (2009). Nesting habits and nest structure of stingless bee, *Trigona iridipennis* Smith at Dharwad, Karnataka. *Karnataka J. Agric. Sci., 22(2):*, (310-313).

Darmawan, S. (n.d.). Diversifikasi Produk Trigona sp. Di Lombok. *Badan Litbang Kehutanan, Kementrian Kehutanan.*

Dollin, A. (1996). Nests of Australian stingless bees. *Australian Native Bee Research Center, North Richmond, NSW.*, 14 pp.

Dollin, A. E., Dollin, L. J., & Rasmussen, C. (2015). Australian and New Guinean Stingless Bees of the Genus Austroplebeia Moure (Hymenoptera: Apidae)—a revision. *Zootaxa.*

Dollin, A. E., Dollin, L. J., & Sakagami, S. F. (1997). Australian Stingless Bees of the Genus Trigona (Hymenoptera: Apidae). *CSIRO Australia Invertebrate Taxonomy.*

Engel, M. S., & Rasmussen, C. (2017). A new subgenus of Heterotrigona from New Guinea (Hymenoptera: Apidae). *Journal of Melittology.*

Engel, M. S., Michener, C. D., & Boontop, Y. (2017). Notes on Southeast Asian Stingless Bees of the Genus Tetragonula (Hymenoptera: Apidae), with the Description of a New Species from Thailand. *American Museum Novitates*.

Fatoni, A. (2008). Pengaruh propolis Trigona spp. asal Bukittinggi terhadap beberapa bakteri usus halus sapi dan penelusuran komponen aktifnya [Tesis]. Bogor (ID):. *Program Pascasarjana, Institut Pertanian Bogor, Bogor.*

Freise, H. (1898). Die Trigona-arten Australiens. *Termeszetrajzi Fuzetek 21:*, 427-431.

Heard, T. (2016). The Australian Native Bee Book. *ISBN 978-0-646-93997.*

Inoue, T., Salmah, S., Abbas, I., & Yusuf, E. (1985). 1985. Foraging Behavior of Individual Workers and Foraging Dynamics of Colonies of Three Sumatran Stingless bees. *Res. Popul. Ecol. 27:*, 373-392.

Jalil, A. (2016). *World Meliponine Etymology of Taxanomic Nomenclature.* Akademi Kelulut Malaysia.

Jalil, A. H. (2018). *Southern Luzon Meliponiculture Tour.* Luzon, The Philippines: Akademu Kelulut Malaysia.

Jalil, A. H. (2019). *Visayas Meliponiculture Tour.* Cebu, The Philippines: Akademi Kelulut Malaysia.

Jalil, A. H., & Roubik(ed), D. W. (2019). *Filipino Meliponiculture and Beyond inc. MIMAROPA Tours.* Mindoro, The Philippines: Akademu Kelulut Malaysia.

Jalil, A., & Roubik(ed), D. (2016). *Handbook of Meliponiculture.* Akademi Kelulut Malaysia.

Jalil, A., & Roubik(ed), D. (2018). Handbook of Meliponiculture Vol 2. *Akademi Kelulut malaysia.*

Jongjitvimol, T., Boontawon, K., & Deowanish, W. W. (2005). Nest Dispersion of a Stingless Bee Species, Trigona collina Smith, 1857 (Apidae, Meliponinae) in a Mixed Deciduous Forest in Thailand. *The Natural History Journal of Chulalongkorn University 5(2): 69-71.*

Kumar, M., Singh, A., & Alagumuthu, G. (2012). Traditional Beekeeping of Stingless Bee (Trigona asp) by Kani Tribes of Western Ghats, Tamil Nadu, India. *Indian Journal of Traditional Knowledge. 11(2):* 342-345.

Lamerkabel, J. S., & et al. (2017). Tabiat Bersarang Dua Spesis Stingless Bee Koloni Di Pulau Ambon Muluku. *Seminar Nasional Perlebahan.* Bogor: IPB.

Michener, C. D. (1961). Observations on the Nests and Behavior of Trigona in Australia and New Guinea (Hymenoptera, Apidae). *American Museum Novitates.*

Michener, C. D. (2007). The Bees of the World 2nd ed. *The John Hopkins university press.*

Moure's Bee Catalogue. (23 Jul, 2008). *Bee Catalogue.* (Universidade Federal do Paraná, UFPR, Centro de Referência em Informação Ambiental, CRIA) Retrieved 17 Jun, 2013, from Moure's Bee Catalogue: http://moure.cria.org.br/catalogue

Nogueira-Neto, P. (1997). Life and Creating Indigenous bees stingless, São Paulo: *Editora Nogueirapis,* 1997.

Pham, H., Phan, T., & & Ha, T. (2021). Studying the Biological Characteristics of the Stingless Bee *Tetrigonilla Collina* Smith 1857 (Apidae: Melipona) in Hanoi, Vietnam. VJAS 2021; 4(2): 1067-1076 .

Rahman1, A., Das, P. K., Rajkumari, P., Saikia, J., & Sharmah, D. (2015). Stingless Bees (Hymenoptera:Apidae: Meliponini): Diversity and Distribution in India. *International Journal of Science and Research (IJSR)*, Volume 4 Issue 1.

Rasmussen. (2008). Catalogue of the Indo-Malayan/Australasian stingless bees (Hymenoptera: Apidae: Meliponini). DOI: https://doi.org/10.11646/zootaxa.1935.1.1

Rasmussen, C. (2013). Stingless bees (Hymenoptera: Apidae: Meliponini) of the Indian subcontinent: Diversity, taxonomy and current status of knowledge. *Zootaxa 3647 (3): 401–428.*

Rasmussen, C., & Camargo, J. M. (2008). A molecular phylogeny and the evolution of nest architecture and behavior in Trigona s.s. (Hymenoptera: Apidae: Meliponini). *Apidologie 39.*

Rasmussen, C., Thomas, J. C., & Engel, M. S. (2017). A New Genus of Eastern Hemisphere Stingless Bees (Hymenoptera: Apidae), with a Key to the Supraspecific Groups of Indomalayan and Australasian Meliponini. *American Museum Novitates.*

Rattanawannee, A., & Chanchao, C. (n.d.). Bee Diversity in Thailand and the Applications of Bee Products. *Department of Biology, Faculty of Science, Chulalongkorn University, Bangkok, Thailand.*

Riendriasari, S. D. (2014). Budidaya Dan Produk Perlebahan Trigona Spp Di Lombok, Nusa Tenggara Barat. *Prosiding Seminar Nasional* (pp. 213 -221). Jogja: University Club Universitas Gadjah Mada,.

Roubik, D. (2006). Stingless bee nesting biology. *Apidologie 37: 124-143*

Sakagami, S. F. (1978). *Tetragonula* Stingless Bees of the Continental Asia and Sri Lanka (Hymenoptera, Apidae). *Jour. Fac. Sci. Hokkaido Univ. Ser. VZ, 2001. 21(2).*

Sakagami, S. F., & Inoue, T. (1985). Taxonomic Notes on Three Bicolorous *Tetragonula* Stingless Bees in Southeast Asia. *Kontyii, Tokyo, 53 (1): 174-189.*

Sakagami, S. F., Inoue, T., & Salmah, S. (1990). Stingless Bees of Central Sumatra. *Sapporo; Hokkaido University Press.*

Sakagami, S. F., Inoue, T., Yamane, S., & Salmah, S. (1983). Nest architecture and colony composition of the Sumatran stingless bee *Trigona (Tetragona) laeviceps. Kontyu, 51(1),* 100-111.

Schwarz, H. F. (1937). Bornean Stingless Bees Of The Genus *Trigona. Bulletin American Museum of Natural History.*

Schwarz, H. F. (1939). The Indo Malayan Species of *Trigona. Bulletin of AMNH.*

Senckenberg Biographies. (11 December 2013). *World of Biodiversity*, 6. (Senckenberg) Retrieved 20 April 2014, from Biographies of entomologists of the world: http://sdei.senckenberg.de/biographies/

Sommeijer, M. (1999). Beekeeping with stingless bees: a new type of hive. *Bee World 80(2):* 70-79.

Starr, C. K., & Sakagami, S. F. (1987). An Extraordinary Concentration of Stingless Bee in the Philippines, with Notes on Nest Architecture (Hymenoptera: Apidae: *Trigona* spp). *Insectes Sociaux 34(2):96-107.*

Thummajitsakul, S., Klinbunga, S., Smith, D., & Sittipraneed, S. (2008). Genetic diversity and population structure of *Trigona pagdeni* Schwarz in Thailand. *Apidologie 39*, 446–455.

Vasanthakumar, S., Srinivasan, M. R., & Thakur, R. K. (2017). Pollination behaviour of stingless bees *Tetragonula iridipennis* Smith on mango inflorescence in south India. *AKM- IMS3C*.

Viraktamath, S., & Jose, K. S. (2017). Two New Species Of *Lisotrigona* Moure (Hymenoptera:Apidae: Meliponini) From India With Notes On Nest Structure. *The Bioscan 12(1): 21-28, 2017*, 21-28.

Vit, P., Pedro, S., & Roubik, D. (2018). Pot-Pollen in Stingless Bee Melittology. *Springer International Publishing AG*. DO - 10.1007/978-3-319-61839-5.

Vit, P., Roubik, D. W., & Pedro, S. M. (2013). Pot Honey - A Legacy of Stingless Bees. *Springer*, DO - 10.1007/978-1-4614-4960-7.

Wikipedia. (20 July 2013). *Frederick Smith*. Retrieved 21 April 2014, from http://en.wikipedia.org/wiki/Frederick_Smith_(entomologist)

Wikipedia. (28 December 2013). *Theodore Dru Alison Cockerell* . Retrieved 20 April 2014, from http://en.wikipedia.org/wiki/Theodore_Dru_Alison_Cockerell

Wikipedia. (7 February 2014). *Adolpho Ducke*. Retrieved 20 April 2014, from http://en.wikipedia.org/wiki/Adolpho_Ducke

Wikipedia. (18 March 2014). *Johann Karl Wilhelm Illiger*. Retrieved 16 May 2014, from http://en.wikipedia.org/wiki/Johann_Karl_Wilhelm_Illiger

Wikipedia. (15 March 2014). *Louis Jurine*. Retrieved 7 May 2014, from http://en.wikipedia.org/wiki/Louis_Jurine: http://en.wikipedia.org/wiki/Louis_Jurine

Xiong, C. S. (2015). A Conservation Assessment of Stingless Bees (Apidae: Meliponini) in Singapore. *Cohort AY2014/2015 S1*.

Index

APPENDIX A

Synonymic Checklist of Indomalayan-Australasian Meliponini Based on Synonymy Checklist in Rasmussen, Thomas & Engel, 2017.

The following list summarizes the currently recognized species within the genera and subgenera of stingless bees occurring in the Indomalayan and Australasian regions. Currently, valid names are indicated in boldface, with known synonyms listed, and asterisks identify the type species for the genus-group taxa.

Genus	Sub Genus	Valid Species	Synonymy
Austroplebeia Moure 1961		*Austroplebeia* ***australis*** (Friese, 1898)	*Trigona cincta percincta* Cockerell, 1929a
			Trigona cockerelli Rayment, 1930
			Trigona cockerelli ornata Rayment, 1932
			Trigona websteri Rayment, 1932
		Austroplebeia ***cassiae*** (Cockerell, 1910) *	*Trigona symei* Rayment, 1932
		Austroplebeia ***cincta*** (Mocsáry in Friese, 1898)	
		Austroplebeia ***essingtoni*** (Cockerell, 1905)	
		Austroplebeia ***magna*** Dollin, Dollin, and Rasmussen, 2015	
Geniotrigona Moure, 1961		*Geniotrigona* ***lacteifasciata*** (Cameron, 1902)	*Trigona borneënsis* Friese, 1933a
		Geniotrigona ***thoracica*** (Smith, 1857) *	*Trigona ambusta* Cockerell, 1918
Heterotrigona Schwarz, 1939	*Heterotrigona* Schwarz, 1939	*Heterotrigona (Heterotrigona)* ***bakeri*** (Cockerell, 1919b)	
		Heterotrigona (Heterotrigona) ***erythrogastra*** (Cameron, 1902)	*Trigona luteiventris* Friese, 1908 [1909]
			Trigona sandacana Cockerell, 1919a
		Heterotrigona (Heterotrigona) ***itama*** (Cockerell, 1918) *	*Trigona breviceps* Cockerell, 1919a
	Platytrigona Moure, 1961	*Heterotrigona (Platytrigona)* ***flaviventris*** (Friese, 1908 [1909])	
		Heterotrigona (Platytrigona) ***hobbyi*** (Schwarz, 1937)	
		Heterotrigona (Platytrigona) ***keyensis*** (Friese, 1901)	
		Heterotrigona (Platytrigona) ***lamingtonia*** (Cockerell, 1929a)	
		Heterotrigona (Platytrigona) ***planifrons*** (Smith, 1865) *	
	Sahulotrigona Engel and Rasmussen, 2017	*Heterotrigona (Sahulotrigona)* ***atricornis*** (Smith, 1865)	

		Heterotrigona (Sahulotrigona) ***paradisaea*** Engel and Rasmussen, 2017*	*Trigona (Pacificotrigona) okazawai* Tadauchi et al., 1998, nomen nudum
			Trigona (Pacificotrigona) okazawai irianensis Tadauchi et al., 1998, nomen nudum
	Sundatrigona Inoue and Sakagami, 1993	*Heterotrigona (Sundatrigona)* ***lieftincki*** (Sakagami and Inoue, 1987)	
		Heterotrigona (Sundatrigona) ***moorei*** (Schwarz, 1937) *	*Trigona (Tetragona) matsumurai* Sakagami, 1959
Homotrigona Moure, 1961	*Homotrigona* Moure, 1961	*Homotrigona (Homotrigona)* ***aliceae*** (Cockerell, 1929b)	*Trigona kusutkana* Dover, 1929
		Homotrigona (Homotrigona) ***anamitica*** (Friese, 1909)	*Trigona melanotricha* Cockerell, 1918
		Homotrigona (Homotrigona) ***fimbriata*** (Smith, 1857) *	*Melipona castanea* Bingham, 1903
			Trigona flavistigma Cameron, 1902
			Trigona versicolor Friese, 1908 [1909]
		Homotrigona (Homotrigona) ***lutea*** (Bingham, 1897)	*Trigona ferrea* Cockerell, 1929b
	Lophotrigona Moure, 1961	*Homotrigona (Lophotrigona)* ***canifrons*** (Smith, 1857) *, new combination	*Trigona busara* Cockerell, 1918
	Odontotrigona Moure, 1961	*Homotrigona (Odontotrigona)* ***haematoptera*** (Cockerell, 1919a) *, new combination	*Trigona haematoptera variety dulitae* Schwarz, 1937
	Tetrigona Moure, 1961	*Homotrigona (Tetrigona)* ***apicalis*** (Smith, 1857) *, new combination	*Trigona hemileuca* Cockerell, 1929b
			Trigona sericea Friese, 1933b
		Homotrigona (Tetrigona) ***binghami*** (Schwarz, 1937), new combination	
		Homotrigona (Tetrigona) ***melanoleuca*** (Cockerell, 1929b), new combination	
		Homotrigona (Tetrigona) ***peninsularis*** (Cockerell, 1927), new combination	
		Homotrigona (Tetrigona) ***vidua*** (Lepeletier de Saint Fargeau, 1836), new combination	
Lepidotrigona Schwarz, 1939		*Lepidotrigona* ***arcifera*** (Cockerell, 1929c)	*Melipona tunneli* Pugh, 1947, nomen dubium et inquirendum
		Lepidotrigona ***doipaensis*** (Schwarz, 1939)	
		Lepidotrigona ***flavibasis*** (Cockerell, 1929c)	

		Lepidotrigona ***hoozana*** (Strand, 1913)	
		Lepidotrigona ***javanica*** (Gribodo, 1891)	
		Lepidotrigona ***latebalteata*** (Cameron, 1902)	
		Lepidotrigona ***latipes*** (Friese, 1900)	
		Lepidotrigona ***nitidiventris*** (Smith, 1857) *	*Trigona fulvopilosella* Cameron, 1908
		Lepidotrigona ***palavanica*** (Cockerell, 1915)	
		Lepidotrigona ***terminata*** (Smith, 1878)	*Trigona fulvomarginata* Cockerell, 1919b
		Lepidotrigona ***trochanterica*** (Cockerell, 1920)	
		Lepidotrigona ***ventralis*** (Smith, 1857)	*Trigona (Lepidotrigona) ventralis form nigribasis* Sakagami, 1975, nomen invalidum
Lisotrigona Moure, 1961		*Lisotrigona* ***cacciae*** (Nurse, 1907) *	*Trigona scintillans* Cockerell, 1920
			Lisotrigona mohandasi Jobiraj and Narendran, 2004, new synonym
			Lisotrigona chandrai Viraktamath and Sajan Jose, 2017, new synonym
			Lisotrigona revanai Viraktamath and Sajan Jose, 2017, new synonym
		Lisotrigona ***carpenteri*** Engel, 2000	
		Lisotrigona ***furva*** Engel, 2000	
Papuatrigona Michener and Sakagami, 1990		*Papuatrigona* ***genalis*** (Friese, 1908 [1909]) *	
Pariotrigona Moure, 1961		*Pariotrigona* ***pendleburyi*** *(Schwarz, 1939)* *	*Trigona (Hypotrigona) pendleburyi variety klossi* Schwarz, 1939
Tetragonula Moure, 1961	*Tetragonilla* Moure, 1961	*Tetragonula (Tetragonilla)* ***atripes*** (Smith, 1857) *	
		Tetragonula (Tetragonilla) ***collina*** (Smith, 1857)	*Trigona cambodiensis* Cockerell, 1926
		Tetragonula (Tetragonilla) ***fuscibasis*** (Cockerell, 1920)	
		Tetragonula (Tetragonilla) ***rufibasalis*** (Cockerell, 1918)	
	Tetragonula Moure, 1961	*Tetragonula (Tetragonula)* ***bengalensis*** (Cameron, 1897)	

		Tetragonula (Tetragonula) ***biroi*** (Friese, 1898)	
		Tetragonula (Tetragonula) ***carbonaria*** (Smith, 1854)	*Trigona angophorae* Cockerell, 1912
		Tetragonula (Tetragonula) ***clypearis*** (Friese, 1909)	*Trigona wybenica* Cockerell, 1929d
		Tetragonula (Tetragonula) ***dapitanensis*** (Cockerell, 1925)	
		Tetragonula (Tetragonula) ***davenporti*** (Franck in Franck et al., 2004)	
		Tetragonula (Tetragonula) ***drescheri*** (Schwarz, 1939)	
		Tetragonula (Tetragonula) ***fuscobalteata*** (Cameron, 1908)	*Trigona erythrostoma* Cameron, 1908
			Trigona pallidistigma Cameron, 1908
			Trigona atomella Cockerell, 1919a
			Trigona brevis Friese, 1925, nomen nudum
			Trigona pfeifferi Friese, 1925
			Trigona pygmaea Friese, 1933b
		Tetragonula (Tetragonula) ***geissleri*** (Cockerell, 1918)	*Trigona confusella* Cockerell, 1919a
		Tetragonula (Tetragonula) ***gressitti*** (Sakagami, 1978)	
		Tetragonula (Tetragonula) ***hirashimai*** (Sakagami, 1978)	
		Tetragonula (Tetragonula) ***hockingsi*** (Cockerell, 1929e)	
		Tetragonula (Tetragonula) ***iridipennis*** (Smith, 1854) *	
		Tetragonula (Tetragonula) ***laeviceps*** (Smith, 1857)	
		Tetragonula (Tetragonula) ***malaipanae*** Engel, Michener, and Boontop, 2017	
		Tetragonula (Tetragonula) ***melanocephala*** (Gribodo, 1893)	*Trigona testaceinerva* Cameron, 1908
		Tetragonula (Tetragonula) ***melina*** *(Gribodo, 1893)*	
		Tetragonula (Tetragonula) ***mellipes*** *(Friese, 1898)*	
		Tetragonula (Tetragonula) ***minangkabau*** (Sakagami and Inoue, 1985)	*Trigona minangkabu form darek* Sakagami and Inoue, 1985, nomen invalidum et nudum
		Tetragonula (Tetragonula) ***minor*** (Sakagami, 1978)	

		Tetragonula (Tetragonula) ***pagdeni*** (Schwarz, 1939)	
		Tetragonula (Tetragonula) ***pagdeniformis*** (Sakagami, 1978)	
		Tetragonula (Tetragonula) ***penangensis*** (Cockerell, 1919b)	
		Tetragonula (Tetragonula) ***praeterita*** (Walker, 1860)	
		Tetragonula (Tetragonula) ***reepeni*** (Friese, 1918)	*Trigona latigenalis* Sakagami, 1978
		Tetragonula (Tetragonula) ***ruficornis*** (Smith in Horne and Smith, 1870)	*Melipona smithii* Bingham, 1897, nomen vanum
		Tetragonula (Tetragonula) ***sapiens*** (Cockerell, 1911)	
		Tetragonula (Tetragonula) sarawakensis (Schwarz, 1937)	
		Tetragonula (Tetragonula) ***sirindhornae*** *(Michener and Boongird, 2004)*	
		Tetragonula (Tetragonula) ***testaceitarsis*** *(Cameron, 1901)*	*Trigona testaceicornis* Cameron, 1901, lapsis calami
			Trigona valdezi Cockerell, 1918
		Tetragonula (Tetragonula) ***zucchii*** (Sakagami, 1978)	
Wallacetrigona Engel and Rasmussen, 2017		*Wallacetrigona* ***incisa*** (Sakagami and Inoue, 1989) *, new combination	
Nomina dubia and unassociated nomina nuda		*Melipona basimaculata* Bingham, 1903, nomen dubium et inquirendum	
		Melipona khasiana Pugh, 1947, nomen nudum	
		Melipona pulla Illiger, 1806, nomen nudum	
		Melipona terrestris Pugh, 1947, nomen nudum	
		Trigona bismarkiana Tadauchi et al., 1998, nomen nudum	
		Trigona pallidicincta Cockerell, 1918, nomen dubium et inquirendum	
		Trigona wallacei Tadauchi et al., 1998, nomen nudum (a species of Tetragonula s.str.)	

Appendix B - Regional Distribution of Meliponiculture of *Tetragonula* spp.

Distribution of Meliponines in Thailand, Cambodia & Vietnam

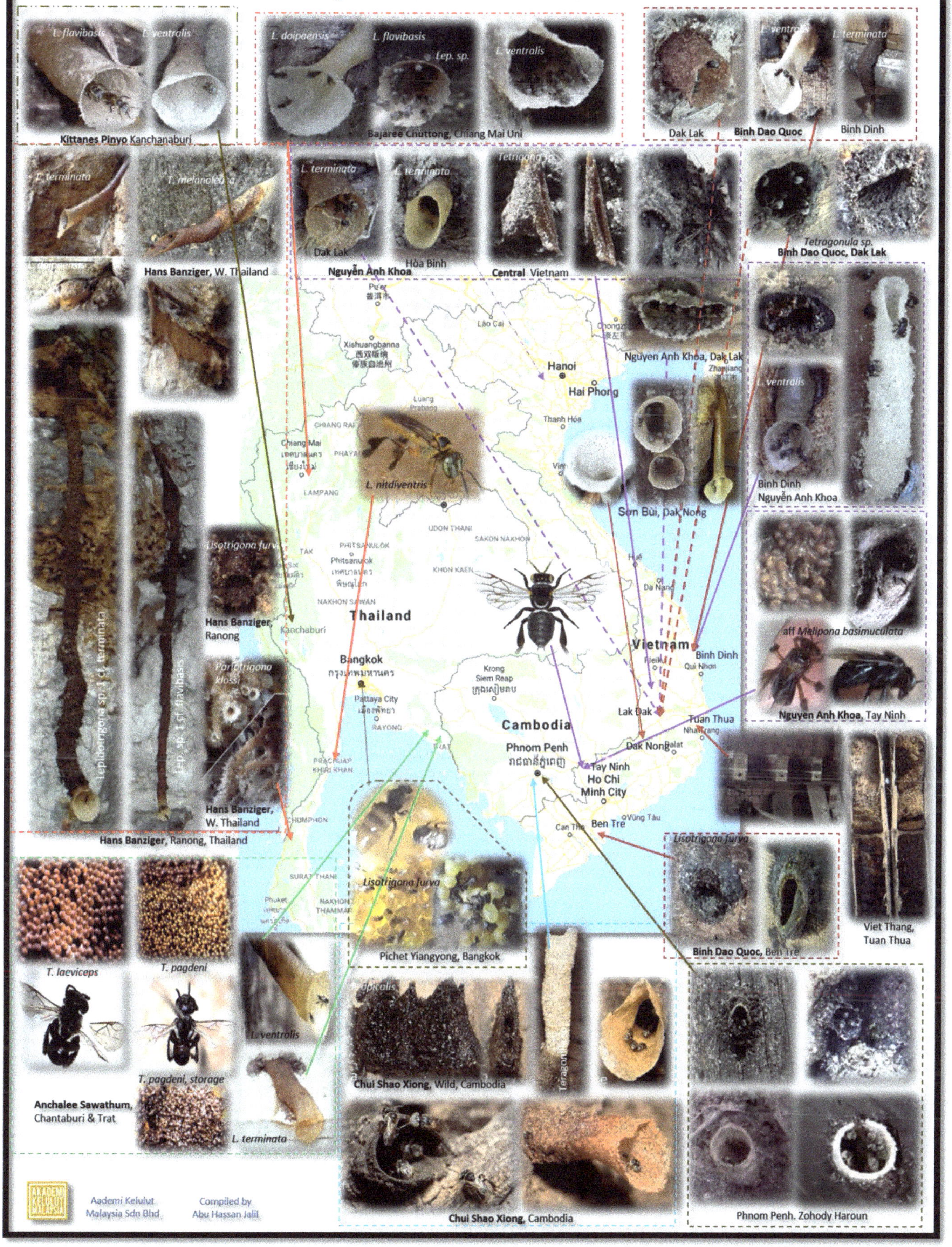

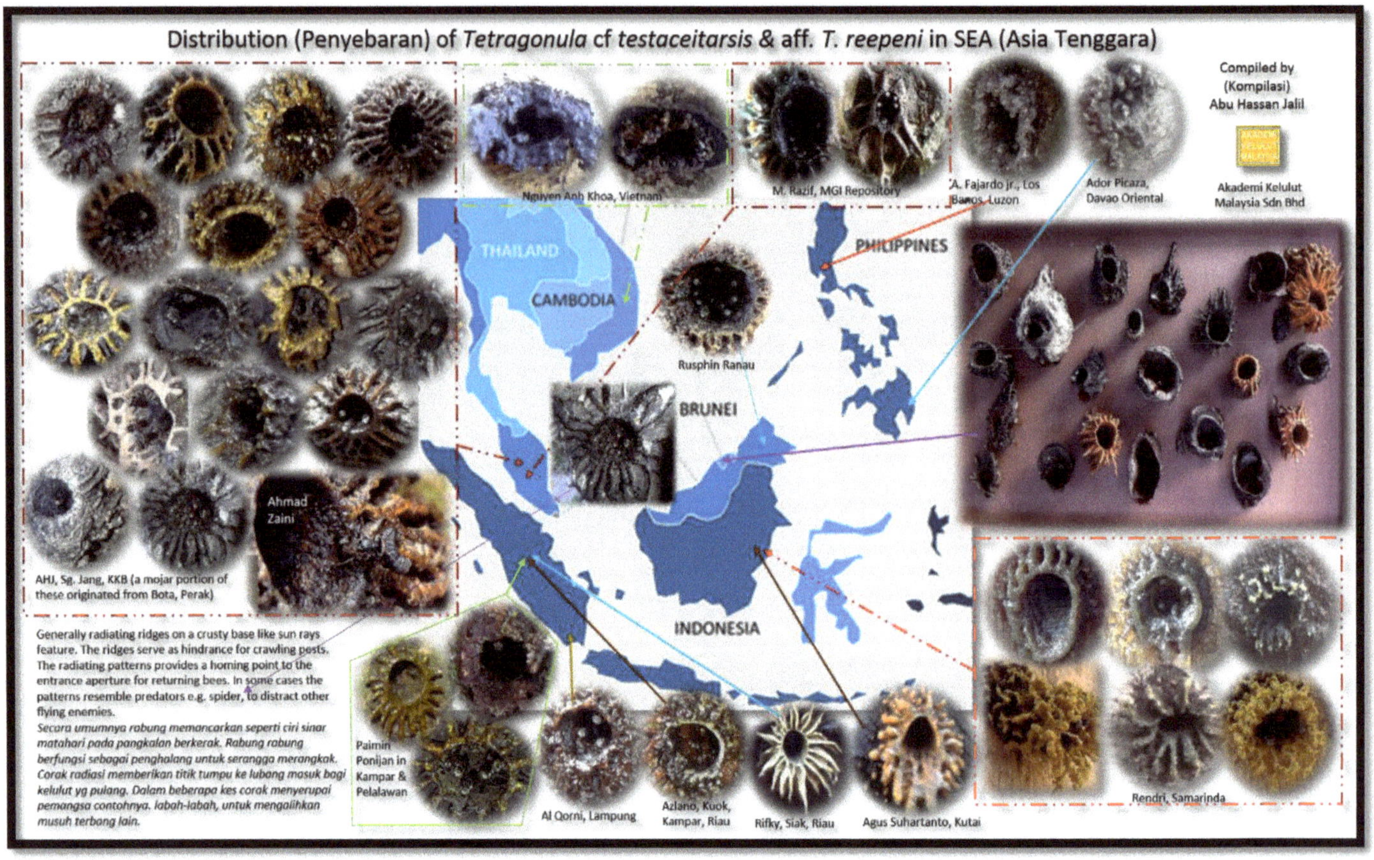
Distribution (Penyebaran) of *Tetragonula* cf *testaceitarsis* & aff. *T. reepeni* in SEA (Asia Tenggara)
Compiled by (Kompilasi) Abu Hassan Jalil
Akademi Kelulut Malaysia Sdn Bhd
Nguyen Anh Khoa, Vietnam
M. Razif, MGI Repository
A. Fajardo jr., Los Banos, Luzon
Ador Picaza, Davao Oriental
THAILAND
CAMBODIA
PHILIPPINES
Rusphin Ranau
BRUNEI
INDONESIA
Ahmad Zaini
AHJ, Sg. Jang, KKB (a mojar portion of these originated from Bota, Perak)
Generally radiating ridges on a crusty base like sun rays feature. The ridges serve as hindrance for crawling pests. The radiating patterns provides a homing point to the entrance aperture for returning bees. In some cases the patterns resemble predators e.g. spider, to distract other flying enemies.
Secara umumnya rabung memancarkan seperti ciri sinar matahari pada pangkalan berkerak. Rabung rabung berfungsi sebagai penghalang untuk serangga merangkak. Corak radiasi memberikan titik tumpu ke lubang masuk bagi kelulut yg pulang. Dalam beberapa kes corak menyerupai pemangsa contohnya. labah-labah, untuk mengalihkan musuh terbang lain.
Paimin Ponijan in Kampar & Pelalawan
Al Qorni, Lampung
Azlano, Kuok, Kampar, Riau
Rifky, Siak, Riau
Agus Suhartanto, Kutai
Rendri, Samarinda

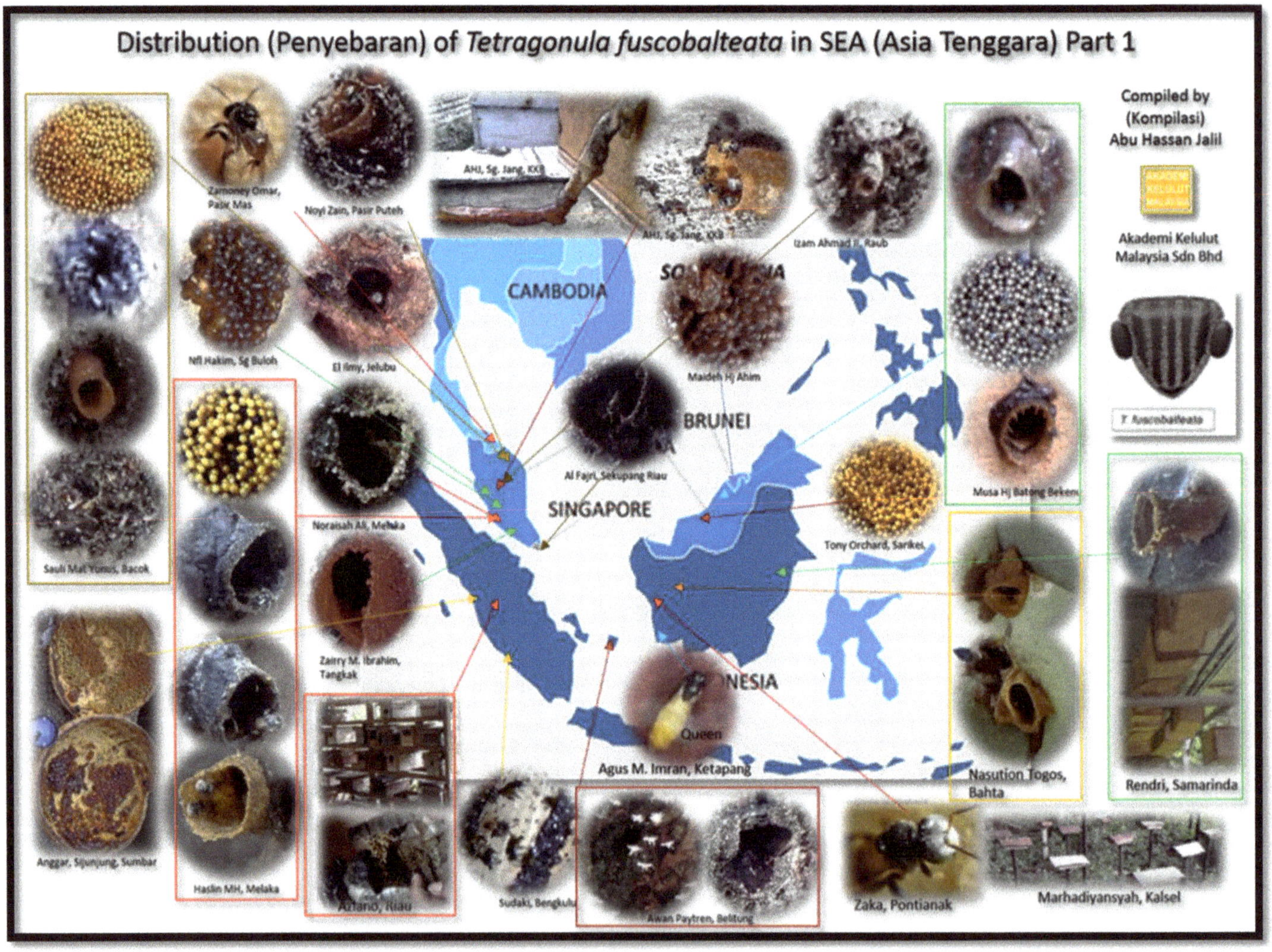
Distribution (Penyebaran) of *Tetragonula fuscobalteata* in SEA (Asia Tenggara) Part 1
Compiled by (Kompilasi) Abu Hassan Jalil
Akademi Kelulut Malaysia Sdn Bhd
Zamoney Omar, Pasir Mas
Noyi Zain, Pasir Puteh
AHJ, Sg. Jang, KKB
AHJ, Sg. Jang, KKB
Izam Ahmad II, Raub
Nfl Hakim, Sg Buloh
El Ilmy, Jelubu
CAMBODIA
Maideh Hj Ahim
BRUNEI
Al Fajri, Sekupang Riau
SINGAPORE
Musa Hj Batong Bekenu
Tony Orchard, Sarikei,
Noraisah Ali, Melaka
Sauli Mat Yunus, Bacok
Zairry M. Ibrahim, Tangkak
Queen
Agus M. Imran, Ketapang
Nasution Togos, Bahta
Rendri, Samarinda
Anggar, Sijunjung, Sumbar
Haslin MH, Melaka
Azlano, Riau
Sudaki, Bengkulu
Awan Paytren, Belitung
Zaka, Pontianak
Marhadiyansyah, Kalsel

Distribution (Penyebaran) of *Tetragonula fuscobalteata* in SEA (Asia Tenggara) Part 2
Compiled by
(Kompilasi)
Abu Hassan Jalil
Akademi Kelulut Malaysia Sdn Bhd
AHJ, AKM, Bangi
AHJ, MGI, Bangi
CAMBODIA
Fathur, Sg Ulin, Kalsel
Ibnu Kasir, Binuang
Tasby Amit, Brunei
Khairulnizam, Seremban
Rendri, Samarinda
Zkifli, Kalsel
INDONESIA
Alrahman, Julok, Aceh Timur
Albi, Banjarmasin
Dadareto Teo, Kalsel

Distribution (Penyebaran) of *Tetragonula fuscobalteata* in SEA (Asia Tenggara) Part 3
Nguyen Anh Khoa, Vietnam
Bamboo Wall, Cebu Is., Philippines
Bamboo Wall, Panabo, Davao, Philippines
3 yr old +
2 yr old +
THAILAND
CAMBODIA
SOUTH CHINA SEA
PHILIPPINES
M. Razif, Bangi
Bakhtiar, Aceh Timur
BRUNEI
MALAYSIA
SINGAPORE
Tetragonula fuscobalteata
Usually 1cm or less. Tapers from base to tube opening. 2 to 3 guards at entrance. Tiny bee at 3mm body size. Known to take over from weak Lepidotrigona or other Tetragonula colonies. Photos by Abu Hassan Jalil.
Anggar Dwi, Riau
Queen
Rusphin, Ranau, Sabah
INDONESIA
Pajmin, Kampar
Compiled by
(Kompilasi)
Abu Hassan Jalil
Akademi Kelulut Malaysia Sdn Bhd
4 yr old +
2 yr old +
Zaka Prihatin, Pontianak
Agus Suhartanto, Kutai
Paimin Ponijan, Luwu, Sulsel

Distribution of (Penyebaran) Tetragonula melanocephala, T sirindhornae & T. melina in Maritime SEA (Nusantara Bg 1) Part 1
T. melina
T. sirindhornae
T. melanocephala
T. melina Queen
Zulkepli Hamidon, Jerantut
Ayeim Ghani Maran
En. Adam, Rawang
Azmand Rasland, Genting
Izam Ahmad II, Raub
Compiled by (Kompilasi) Abu Hassan Jalil
Akademi Kelulut Malaysia Sdn Bhd
Badeul Md Shah, Raub
Kamarulazwa Muhammad, Shah Alam
El Ilmy, Jelubu
Norjipin Saidi Kk, Sabah
Daniel Mansali, Panampang
Albert Chia, Kota Belud
Abang Naim, Simunhan, Swk
Jul Kamil, Kucing
Alili azlan Keninagau
Zakaria Said, Ranau
Debra Smith, Tenom, Sabah
Rusphin, Ranau, Sabah
Yank Iyelah, Sambas, Kalbar
Liew Yuk Ming, KK, Sabah

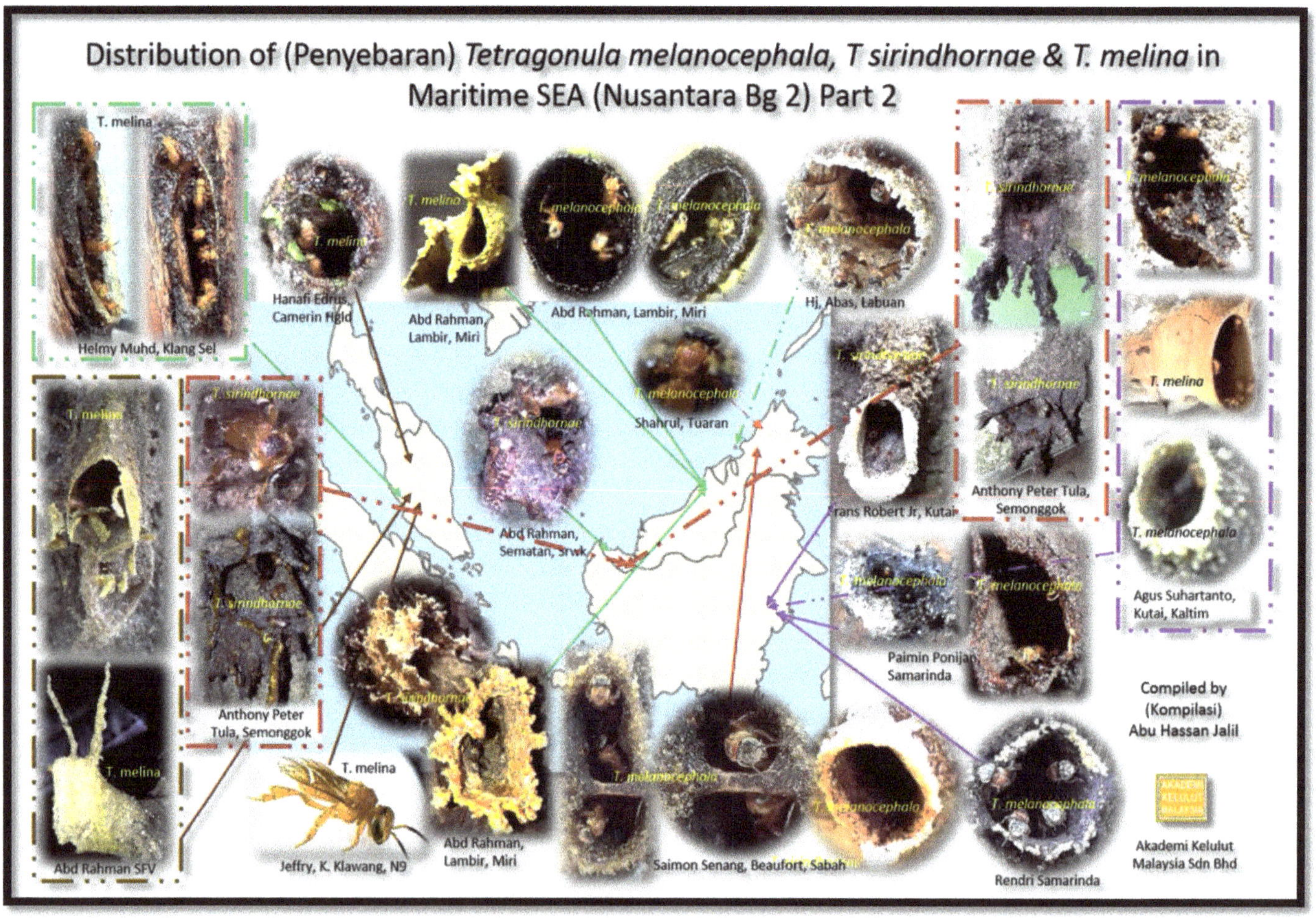

Distribution of (Penyebaran) Tetragonula melanocephala, T sirindhornae & T. melina in Maritime SEA (Nusantara Bg 2) Part 2
T. melina
T. sirindhornae
T. melanocephala
Helmy Muhd, Klang Sel
Hanafi Edris, Camerin Hgld
Abd Rahman, Lambir, Miri
Abd Rahman, Lambir, Miri
Hj. Abas, Labuan
Shahrul, Tuaran
Abd Rahman, Sematan, Srwk
Anthony Peter Tula, Semonggok
Agus Suhartanto, Kutai, Kaltim
Paimin Ponijan Samarinda
Compiled by (Kompilasi) Abu Hassan Jalil
Anthony Peter Tula, Semonggok
Jeffry, K. Klawang, N9
Abd Rahman, Lambir, Miri
Saimon Senang, Beaufort, Sabah
Rendri Samarinda
Abd Rahman SFV
Akademi Kelulut Malaysia Sdn Bhd

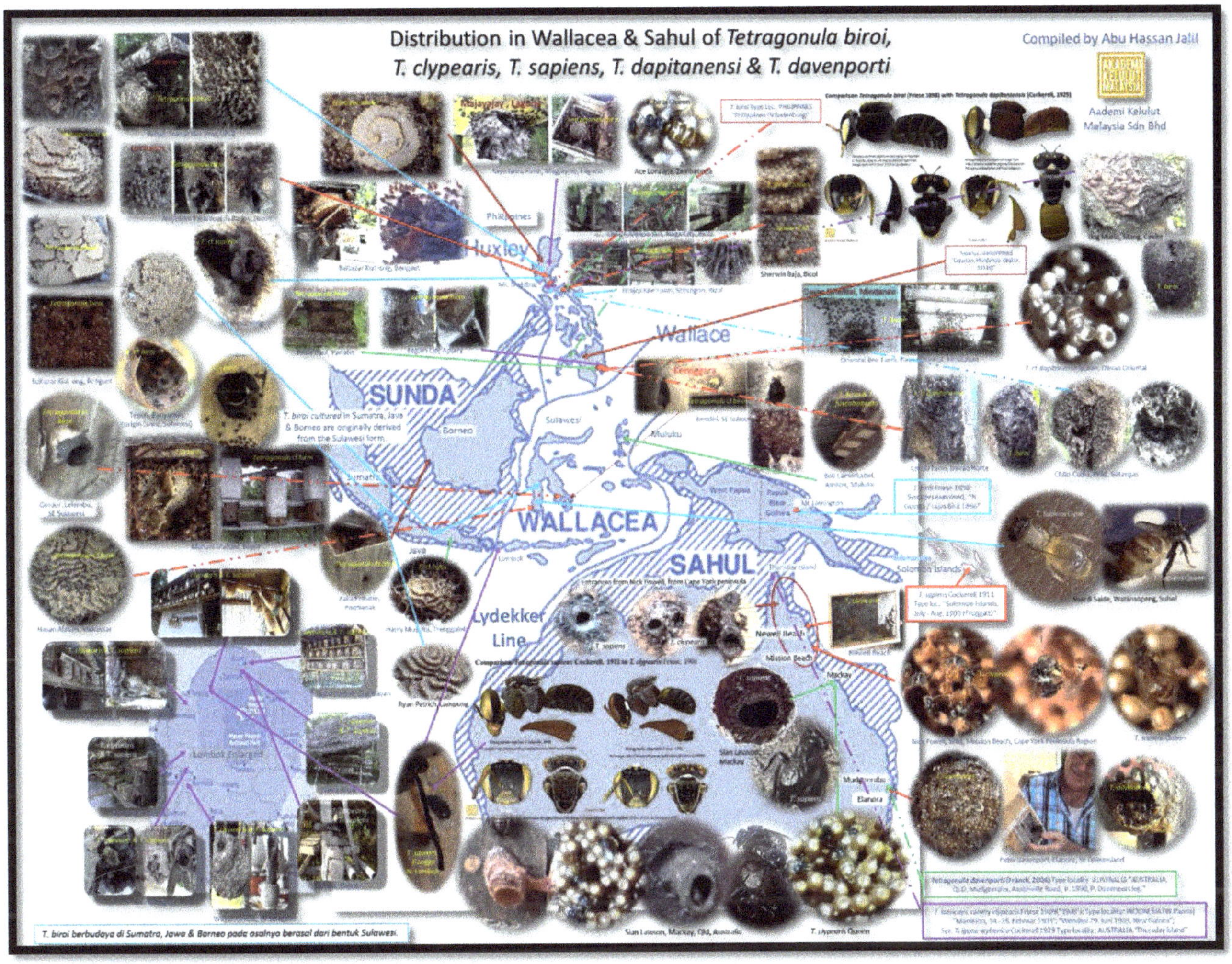

Distribution in Wallacea & Sahul of *Tetragonula biroi*, *T. clypearis*, *T. sapiens*, *T. dapitanensi* & *T. davenporti*
Compiled by Abu Hassan Jalil
Aademi Kelulut Malaysia Sdn Bhd
Philippines
Huxley
Wallace
SUNDA
Borneo
Sulawesi
Maluku
WALLACEA
SAHUL
West Papua
Solomon Islands
Lydekker Line
Java
T. biroi cultured in Sumatra, Java & Borneo are originally derived from the Sulawesi form.
T. biroi berbudaya di Sumatra, Jawa & Borneo pada asalnya berasal dari bentuk Sulawesi.

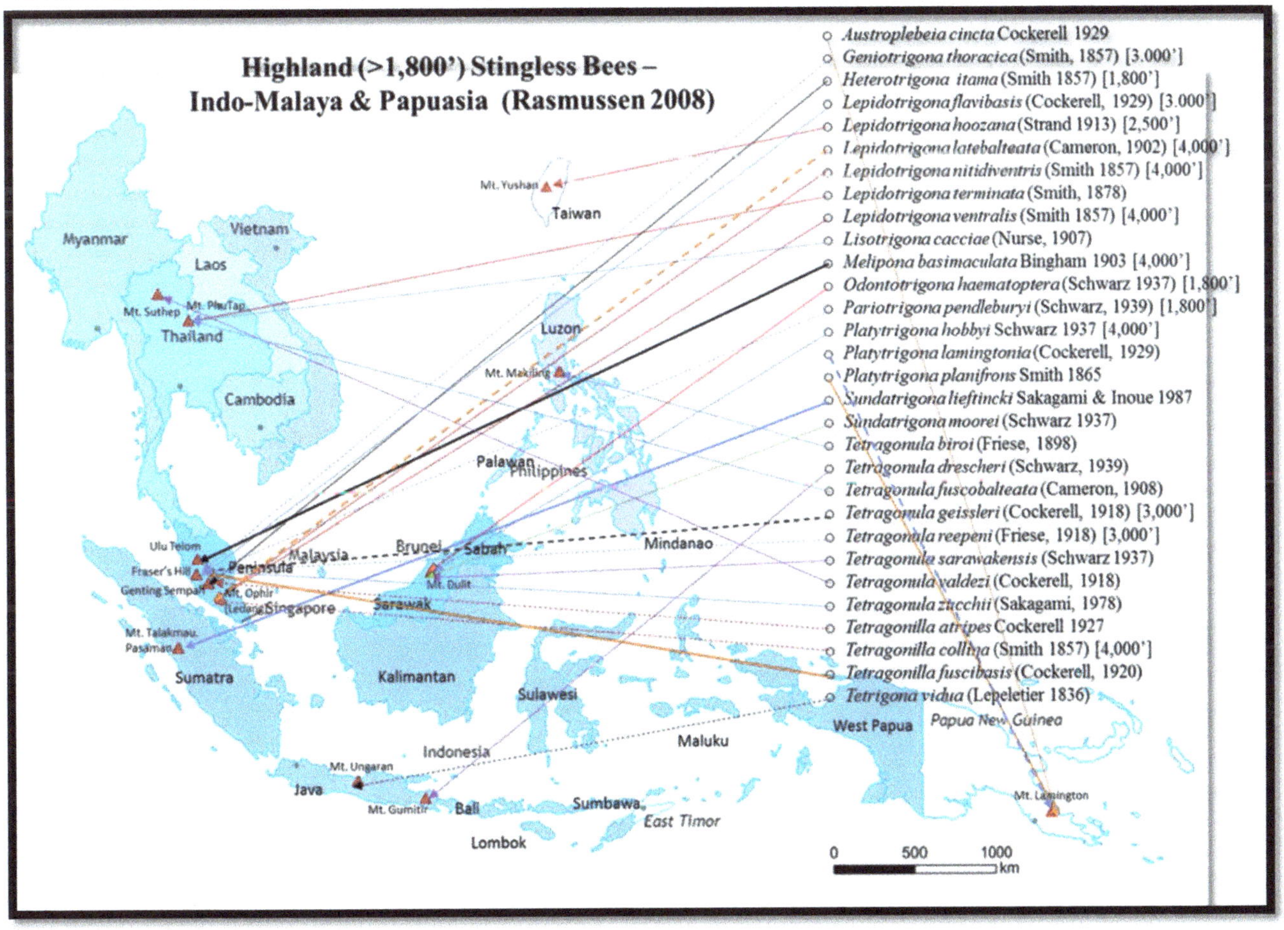

Highland (>1,800') Stingless Bees –
Indo-Malaya & Papuasia (Rasmussen 2008)
Austroplebeia cincta Cockerell 1929
Geniotrigona thoracica (Smith, 1857) [3.000']
Heterotrigona itama (Smith 1857) [1,800']
Lepidotrigona flavibasis (Cockerell, 1929) [3.000']
Lepidotrigona hoozana (Strand 1913) [2,500']
Lepidotrigona latebalteata (Cameron, 1902) [4,000']
Lepidotrigona nitidiventris (Smith 1857) [4,000']
Lepidotrigona terminata (Smith, 1878)
Lepidotrigona ventralis (Smith 1857) [4,000']
Lisotrigona cacciae (Nurse, 1907)
Melipona basimaculata Bingham 1903 [4,000']
Odontotrigona haematoptera (Schwarz 1937) [1,800']
Pariotrigona pendleburyi (Schwarz, 1939) [1,800']
Platytrigona hobbyi Schwarz 1937 [4,000']
Platytrigona lamingtonia (Cockerell, 1929)
Platytrigona planifrons Smith 1865
Sundatrigona lieftincki Sakagami & Inoue 1987
Sundatrigona moorei (Schwarz 1937)
Tetragonula biroi (Friese, 1898)
Tetragonula drescheri (Schwarz, 1939)
Tetragonula fuscobalteata (Cameron, 1908)
Tetragonula geissleri (Cockerell, 1918) [3,000']
Tetragonula reepeni (Friese, 1918) [3,000']
Tetragonula sarawakensis (Schwarz 1937)
Tetragonula valdezi (Cockerell, 1918)
Tetragonula zucchii (Sakagami, 1978)
Tetragonilla atripes Cockerell 1927
Tetragonilla collina (Smith 1857) [4,000']
Tetragonilla fuscibasis (Cockerell, 1920)
Tetrigona vidua (Lepeletier 1836)
Mt. Yushan
Taiwan
Myanmar
Vietnam
Laos
Mt. Suthep
Mt. Phu Tap
Thailand
Cambodia
Luzon
Mt. Makiling
Palawan
Philippines
Mindanao
Ulu Telom
Fraser's Hill
Genting Sempah
Peninsula
Malaysia
Mt. Ophir (Ledang)
Singapore
Brunei
Sabah
Mt. Dulit
Sarawak
Mt. Talakmau, Pasaman
Sumatra
Kalimantan
Sulawesi
West Papua
Papua New Guinea
Indonesia
Maluku
Mt. Ungaran
Java
Mt. Gumitir
Bali
Sumbawa
East Timor
Lombok
Mt. Lamington
0 500 1000 km

Distribution of Meliponiculture in the Philippine Islands Part 1

www.ingramcontent.com/pod-product-compliance
Lightning Source LLC
LaVergne TN
LVHW071201210926
848482LV00014B/702